Bayesian Optimization

Theory and Practice Using Python

Peng Liu

Apress®

Bayesian Optimization: Theory and Practice Using Python

Peng Liu
Singapore, Singapore

ISBN-13 (pbk): 978-1-4842-9062-0
https://doi.org/10.1007/978-1-4842-9063-7

ISBN-13 (electronic): 978-1-4842-9063-7

Managing Director, Apress Media LLC: Welmoed Spahr
Acquisitions Editor: Celestin Suresh John
Development Editor: Laura Berendson
Coordinating Editor: Mark Powers

Cover designed by eStudioCalamar

Cover image by Luemen Rutkowski on Unsplash (www.unsplash.com)

Distributed to the book trade worldwide by Apress Media, LLC, 1 New York Plaza, New York, NY 10004, U.S.A. Phone 1-800-SPRINGER, fax (201) 348-4505, e-mail orders-ny@springer-sbm.com, or visit www.springeronline.com. Apress Media, LLC is a California LLC and the sole member (owner) is Springer Science + Business Media Finance Inc (SSBM Finance Inc). SSBM Finance Inc is a **Delaware** corporation.

For information on translations, please e-mail booktranslations@springernature.com; for reprint, paperback, or audio rights, please e-mail bookpermissions@springernature.com.

Apress titles may be purchased in bulk for academic, corporate, or promotional use. eBook versions and licenses are also available for most titles. For more information, reference our Print and eBook Bulk Sales web page at http://www.apress.com/bulk-sales.

Any source code or other supplementary material referenced by the author in this book is available to readers on GitHub (https://github.com/Apress). For more detailed information, please visit http://www.apress.com/source-code.

Printed on acid-free paper

For my wife Zheng and children Jiaxin, Jiaran, and Jiayu.

Table of Contents

About the Author

Peng Liu is an assistant professor of quantitative finance (practice) at Singapore Management University and an adjunct researcher at the National University of Singapore. He holds a Ph.D. in Statistics from the National University of Singapore and has ten years of working experience as a data scientist across the banking, technology, and hospitality industries.

About the Technical Reviewer

 **Jason Whitehorn** is an experienced entrepreneur and software developer and has helped many companies automate and enhance their business solutions through data synchronization, SaaS architecture, and machine learning. Jason obtained his Bachelor of Science in Computer Science from Arkansas State University, but he traces his passion for development back many years before then, having first taught himself to program BASIC on his family's computer while in middle school. When he's not mentoring and helping his team at work, writing, or pursuing one of his many side-projects, Jason enjoys spending time with his wife and four children and living in the Tulsa, Oklahoma, region. More information about Jason can be found on his website: https://jason.whitehorn.us.

Acknowledgments

This book summarizes my learning journey in Bayesian optimization during my (part-time) Ph.D. study. It started as a personal interest in exploring this area and gradually grew into a book combining theory and practice. For that, I thank my supervisors, Teo Chung Piaw and Chen Ying, for their continued support in my academic career.

Introduction

Bayesian optimization provides a unified framework that solves the problem of sequential decision-making under uncertainty. It includes two key components: a surrogate model approximating the unknown black-box function with uncertainty estimates and an acquisition function that guides the sequential search. This book reviews both components, covering both theoretical introduction and practical implementation in Python, building on top of popular libraries such as GPyTorch and BoTorch. Besides, the book also provides case studies on using Bayesian optimization to seek a simulated function's global optimum or locate the best hyperparameters (e.g., learning rate) when training deep neural networks. The book assumes readers with a minimal understanding of model development and machine learning and targets the following audiences:

- Students in the field of data science, machine learning, or optimization-related fields

- Practitioners such as data scientists, both early and middle in their careers, who build machine learning models with good-performing hyperparameters

- Hobbyists who are interested in Bayesian optimization as a global optimization technique to seek the optimal solution as fast as possible

All source code used in this book can be downloaded from `github.com/apress/ Bayesian-optimization`.

CHAPTER 1

Bayesian Optimization Overview

As the name suggests, Bayesian optimization is an area that studies optimization problems using the Bayesian approach. Optimization aims at locating the optimal objective value (i.e., a global maximum or minimum) of all possible values or the corresponding location of the optimum in the environment (the search domain). The search process starts at a specific initial location and follows a particular policy to iteratively guide the following sampling locations, collect new observations, and refresh the guiding policy.

As shown in Figure 1-1, the overall optimization process consists of repeated interactions between the policy and the environment. The policy is a mapping function that takes in a new input observation (plus historical ones) and outputs the following sampling location in a principled way. Here, we are constantly learning and improving the policy, since a good policy guides our search toward the global optimum more efficiently and effectively. In contrast, a good policy would save the limited sampling budget on promising candidate locations. On the other hand, the environment contains the unknown objective function to be learned by the policy within a specific boundary. When probing the functional value as requested by the policy, the actual observation revealed by the environment to the policy is often corrupted by noise, making learning even more challenging. Thus, Bayesian optimization, a specific approach for *global optimization*, would like to learn a policy that can help us efficiently and effectively navigate to the global optimum of an unknown, noise-corrupted environment as quickly as possible.

P. Liu, *Bayesian Optimization*, https://doi.org/10.1007/978-1-4842-9063-7_1

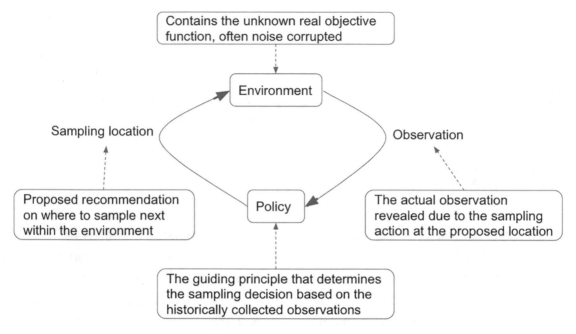

Figure 1-1. *The overall Bayesian optimization process. The policy digests the historical observations and proposes the new sampling location. The environment governs how the (possibly noise-corrupted) observation at the newly proposed location is revealed to the policy. Our goal is to learn an efficient and effective policy that could navigate toward the global optimum as quickly as possible*

Global Optimization

Optimization aims to locate the optimal set of parameters of interest across the whole domain through carefully allocating limited resources. For example, when searching for the car key at home before leaving for work in two minutes, we would naturally start with the most promising place where we would usually put the key. If it is not there, think for a little while about the possible locations and go to the next most promising place. This process iterates until the key is found. In this example, the policy is digesting the available information on previous searches and proposing the following promising location. The environment is the house itself, revealing if the key is placed at the proposed location upon each sampling.

This is considered an easy example since we are familiar with the environment in terms of its structural design. However, imagine locating an item in a totally new environment. The policy would need to account for the uncertainty due to unfamiliarity with the environment while sequentially determining the next sampling location. When the sampling budget is limited, as is often the case in real-life searches in terms of time and resources, the policy needs to argue carefully on the *utility* of each candidate sampling location.

Let us formalize the sequential global optimization using mathematical terms. We are dealing with an unknown scalar-valued objective function f based on a specific domain A. In other words, the unknown subject of interest f is a function that maps a certain sample in A to a real number in $\mathbb{R}$, that is, $f: A \rightarrow \mathbb{R}$. We typically place no specific assumption about the nature of the domain A other than that it should be a bounded, compact, and convex set.

Unless otherwise specified, we focus on the maximization setting instead of minimization since maximizing the objective function is equivalent to minimizing the negated objective, and vice versa. The optimization procedure thus aims at locating the global maximum f^* or its corresponding location x^* in a principled and systematic manner. Mathematically, we wish to locate f^* where

$$f^* = \max_{x \in A} f(x) = f(x^*)$$

Or equivalently, we are interested in its location x^* where

$$x^* = \text{argmax}_{x \in A} f(x)$$

Figure 1-2 provides an example one-dimensional objective function with its global maximum f^* and its location x^* highlighted. The goal of global optimization is thus to systematically reason about a series of sampling decisions within the total search space A, so as to locate the global maximum as fast as possible, that is, sampling as few times as possible.

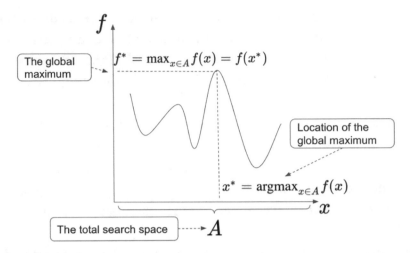

Figure 1-2. *An example objective function with the global maximum and its location marked with star. The goal of global optimization is to systematically reason about a series of sampling decisions so as to locate the global maximum as fast as possible*

Note that this is a *nonconvex function,* as is often the case in real-life functions we are optimizing. A nonconvex function means we could not resort to first-order gradient-based methods to reliably search for the global optimum since it will likely converge to a local optimum. This is also one of the advantages of Bayesian optimization compared with other gradient-based optimization procedures.

The Objective Function

There are different types of objective functions. For example, some functions are wiggly shaped, while others are smooth; some are convex, while others are nonconvex. An objective function is an unknown object to us; the problem would be considered solved if we could access its underlying mathematical form. Many complex functions are almost impossible to be expressed using an explicit expression. For Bayesian optimization, the specific type of objective function typically bears the following attributes:

- We do not have access to the explicit expression of the objective function, making it a "black-box" function. This means that we can only interact with the environment, that is, the objective function, to perform a functional evaluation by sampling at a specific location.

- The returned value by probing at a specific location is often corrupted by noise and does not represent the exact true value of the objective function at that location. Due to the indirect evaluation of its actual value, we need to account for such noise embedded in the actual observations from the environment.

- Each functional evaluation is costly, thus ruling out the option for an exhaustive probing. We need to have a *sample-efficient* method to minimize the number of evaluations of the environment while trying to locate its global optimum. In other words, the optimizer needs to fully utilize the existing observations and systematically reason about the next sampling decision so that the limited resource is well spent on promising locations.

- We do not have access to its gradient. When the functional evaluation is relatively cheap and the functional form is smooth, it would be very convenient to compute the gradient and optimize using the first-order procedure such as gradient descent. Access to the gradient is necessary for us to understand the adjacent curvature of a particular evaluation point. With gradient evaluations, the follow-up direction of travel is easier to determine.

The "black-box" function is challenging to optimize for the preceding reasons. To further elaborate on the possible functional form of the objective, we list three examples in Figure 1-3. On the left is a convex function with only one global minimum; this is considered easy for global optimization. In the middle is a nonconvex function with multiple local optima; it is difficult to ascertain if the current local optimum is also globally optimal. It is also difficult to identify whether this is a flat region vs. a local optimum for a function with a flat region full of saddle points. All three scenarios are in a minimization setting.

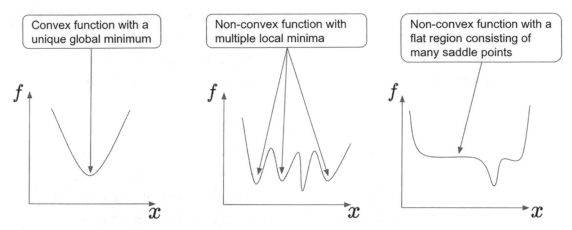

Figure 1-3. *Three possible functional forms. On the left is a convex function whose optimization is easy. In the middle is a nonconvex function with multiple local minima, and on the right is also a nonconvex function with a wide flat region full of saddle points. Optimization for the latter two cases takes a lot more work than for the first case*

Let us look at one example of *hyperparameter tuning* when training machine learning models. A machine learning model is a function that involves a set of parameters to be optimized given the input data. These parameters are automatically tuned via a specific optimization procedure, typically governed by a set of corresponding meta parameters called hyperparameters, which are fixed before the model training starts. For example, when training deep neural networks using the *gradient descent* algorithm, a learning rate that determines the step size of each parameter update needs to be manually selected in advance. If the learning rate is too large, the model may diverge and eventually fails to learn. If the learning rate is too small, the model may converge very slowly as the weights are updated by only a small margin in this iteration. See Figure 1-4 for a visual illustration.

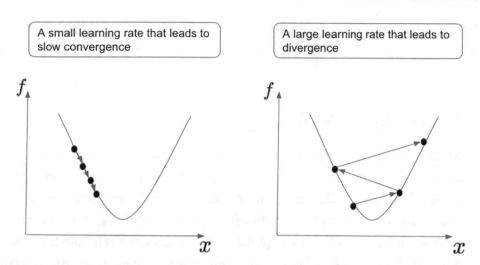

Figure 1-4. *Slow convergence due to a small learning rate on the left and divergence due to a large learning rate on the right*

Choosing a reasonable learning rate as a preset hyperparameter thus plays a critical role in training a good machine learning model. Locating the best learning rate and other hyperparameters is an optimization problem that fits Bayesian optimization. In the case of hyperparameter tuning, evaluating each learning rate is a time-consuming exercise. The objective function would generally be the model's final test set loss (in a minimization setting) upon model convergence. A model needs to be fully trained to obtain a single evaluation, which typically involves hundreds of epochs to reach stable convergence. Here, one epoch is a complete pass of the entire training dataset. The book's last chapter covers a case study on tuning the learning rate using Bayesian optimization.

The functional form of the test set loss or accuracy may also be highly nonconvex and multimodal for the hyperparameters. Upon convergence, it is not easy to know whether we are in a local optimum, a saddle point, or a global optimum. Besides, some hyperparameters may be discrete, such as the number of nodes and layers when training a deep neural network. We could not calculate its gradient in such a case since it requires continuous support in the domain.

The Bayesian optimization approach is designed to tackle all these challenges. It has been shown to deliver good performance in locating the best hyperparameters under a limited budget (i.e., the number of evaluations allowed). It is also widely and successfully used in other fields, such as chemical engineering.

Next, we will delve into the various components of a typical Bayesian optimization setup, including the observation model, the optimization policy, and the Bayesian inference.

The Observation Model

Earlier, we mentioned that a functional evaluation would give an observation about the true objective function, and the observation may likely be different from the true objective value due to noise. The observations gathered for the policy learning would thus be inexact and corrupted by an additional noise term, which is often assumed to be additive. The observation model is an approach to formalize the relationship between the true objective function, the actual observation, and the noise. It governs how the observations would be revealed from the environment to the policy.

Figure 1-5 illustrates a list of observations of the underlying objective function. These observations are dislocated from the objective function due to additive random noises. These additive noises manifest as the vertical shifts between the actual observations and the underlying objective function. Due to these noise-induced deviations inflicted on the observations, we need to account for such uncertainty in the observation model. When learning a policy based on the actual observations, the policy also needs to be robust enough to focus on the objective function's underlying pattern and not be distracted by the noises. The model we use to approximate the objective function, while accounting for uncertainty due to the additive noise, is typically a Gaussian process. We will cover it briefly in this chapter and in more detail in the next chapter.

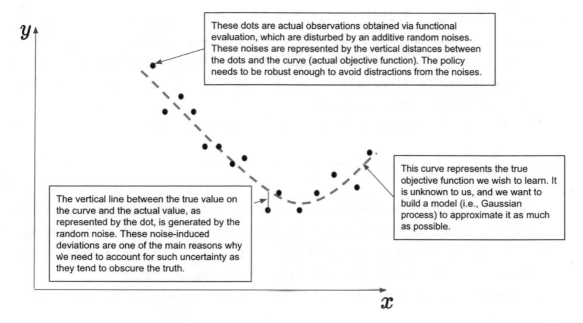

Figure 1-5. *Illustrating the actual observations (in dots) and the underlying objective function (in dashed line). When sampling at a specific location, the observation would be disrupted by an additive noise. The observation model thus determines how the observation would be revealed to the policy, which needs to account for the uncertainty due to noise perturbation*

To make our discussion more precise, let us use $f(x)$ to denote the (unknown) objective function value at location x. We sometimes write $f(x)$ as f for simplicity. We use y to denote the actual observation at location x, which will slightly differ from f due to noise perturbation. We can thus express the observation model, which governs how the policy sees the observation from the environment, as a probability distribution of y based on a specific location x and true function value f:

$$p(y|x, f)$$

Let us assume an additive noise term ε inflicted on f; the actual observation y can thus be expressed as

$$y = f + \varepsilon$$

Here, the noise term ε arises from measurement error or inaccurate statistical approximation, although it may disappear in certain computer simulations. A common practice is to treat the error as a random variable that follows a Gaussian distribution with a zero mean and fixed standard deviation σ, that is, $\varepsilon \sim N(0, \sigma^2)$. Note that it is unnecessary to fix σ across the whole domain A; the Bayesian optimization allows for both homoscedastic noise (i.e., fixed σ across A) and heteroskedastic noise (i.e., different σ that depends on the specific location in A).

Therefore, we can formulate a Gaussian observation model as follows:

$$p(y|x,f,\sigma) = N\left(y; f, \sigma^2\right)$$

This means that for a specific location x, the actual observation y is treated as a random variable that follows a Gaussian/normal distribution with mean f and variance σ^2. Figure 1-6 illustrates an example probability distribution of y centered around f. Note that the variance of the noise is often estimated by sampling a few initial observations and is expected to be small, so that the overall observation model still strongly depends on and stays close to f.

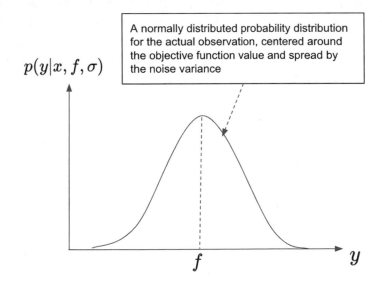

Figure 1-6. *Assuming a normal probability distribution for the actual observation as a random variable. The Gaussian distribution is centered around the objective function f value evaluated at a given location x and spread by the variance of the noise term*

The following section introduces Bayesian statistics to lay the theoretical foundation as we work with probability distributions along the way.

Bayesian Statistics

Bayesian optimization is not a particular algorithm for global optimization; it is a suite of algorithms based on the principles of Bayesian inference. As the optimization proceeds in each iteration, the policy needs to determine the next sampling decision *or* if the current search needs to be terminated. Due to uncertainty in the objective function and the observation model, the policy needs to cater to such uncertainty upon deciding the following sampling location, which bears both an immediate impact on follow-up decisions and a long-term effect on all future decisions. The samples selected thus need to reasonably contribute to the ultimate goal of global optimization and justify the cost incurred due to sampling.

Using Bayesian statistics in optimization paves the way for us to systematically and quantitatively reason about these uncertainties using probabilities. For example, we would place a prior belief about the characteristics of the objective function and quantify its uncertainties by assigning high probability to specific ranges of values and low probability to others. As more observations are collected, the prior belief is gradually updated and calibrated toward the true underlying distribution of the objective function in the form of a posterior distribution.

We now cover the fundamental concepts and tools of Bayesian statistics. Understanding these sections is essential to appreciate the inner workings of Bayesian optimization.

Bayesian Inference

Bayesian inference essentially relies on the Bayesian formula (also called Bayes' rule) to reason about the interactions among three components: the prior distribution $p(\theta)$ where θ represents the parameter of interest, the likelihood $p(\text{data}|\theta)$ given a specific parameter θ, and the posterior distribution $p(\theta|\text{data})$. There is one more component, the evidence of the data $p(\text{data})$, which is often not computable. The Bayesian formula is as follows:

$$p(\theta|\text{data}) = \frac{p(\text{data}|\theta)p(\theta)}{p(\text{data})}$$

Let us look closely at this widely used, arguably the most important formula in Bayesian statistics. Remember that any Bayesian inference procedure aims to derive the posterior distribution $p(\theta|\text{data})$ (or calculate its marginal expectation) for the parameter of interest θ, in the form of a probability density function. For example, we might end up with a continuous posterior distribution as in Figure 1-7, where θ varies from 0 to 1, and all the probabilities (i.e., area under the curve) would sum to 1.

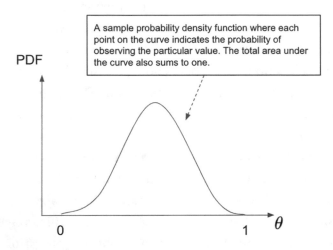

Figure 1-7. *Illustrating a sample (continuous) posterior distribution for the parameter of interest. The specific shape of the curve will change as new data are being collected*

We would need access to three components to obtain the posterior distribution of θ. First, we need to derive the probability of seeing the actual data given our choice of θ, that is, $p(\text{data}|\theta)$. This is also called the likelihood term since we are assessing how likely it is to generate the data after specifying a certain observation model for the data. The likelihood can be calculated based on the assumed observation model for data generation.

The second term $p(\theta)$ represents our prior belief about the distribution of θ without observing any actual data; we encode our pre-experimental knowledge of the parameter θ in this term. For example, $p(\theta)$ could take the form of a uniform distribution that assigns an equal probability to any value between 0 and 1. In other words, all values in this range are equally likely, and this is a common prior belief we would place on θ given that we do not have any information that suggests a preference over specific values. However, as we collect more observations and gather more data, the prior distribution will play a decreasing role, and the subjective belief will gradually reduce in support of the factual evidence in the data. As shown in Figure 1-8, the distribution of θ will progressively approach a normal distribution given that more data is being collected, thus forming a posterior distribution that better approximates the true distribution of θ.

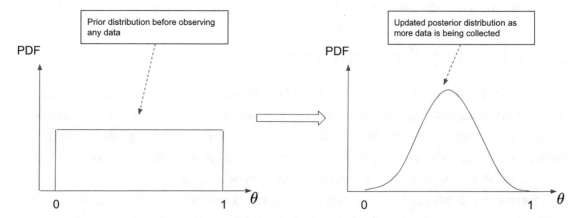

Figure 1-8. *Updating the prior uniform distribution toward a posterior normal distribution as more data is collected. The role of the prior distribution decreases as more data is collected to support the approximation to the true underlying distribution*

The last term is the denominator p(data), also referred to as the evidence, which represents the probability of obtaining the data over all different choices of θ and serves as a normalizing constant independent of θ in Bayes' theorem. This is the most difficult part to compute among all the components since we need to integrate over all possible values of θ by taking an integration. For each given θ, the likelihood is calculated based on the assumed observation model for data generation, which is the same as how the likelihood term is calculated. The difference is that the evidence considers every possible value of θ and weights the resulting likelihood based on the probability of observing a particular θ. Since the evidence is not connected to θ, it is often ignored when analyzing the proportionate change in the posterior. As a result, it focuses only on the likelihood and the prior alone.

A relatively simple case is when the prior $p(\theta)$ and the likelihood p(data$|\theta$) are *conjugate*, making the resulting posterior $p(\theta|$data$)$ analytic and thus easy to work with due to its closed-form expression. Bayesian inference becomes much easier and less restrictive if we can write down the explicit form and generate the exact shape of the posterior $p(\theta|$data$)$ without resorting to sampling methods. The posterior will follow the same distribution as the prior when the prior is conjugate with the likelihood function. One example is when both the prior and the likelihood functions follow a normal distribution, the resulting posterior will also be normally distributed. However, when the prior and the likelihood are not conjugate, we can still get more insight on the posterior distribution via efficient sampling techniques such as *Gibbs sampling*.

Frequentist vs. Bayesian Approach

The Bayesian approach is a systematic way of assigning probabilities to possible values of θ and updating these probabilities based on the observed data. However, sometimes we are only interested in the most probable (expected) value of θ that gives rise to the data we observe. This can be achieved using the *frequentist* approach, treating the parameter of interest (i.e., θ) as a fixed quantity instead of a random variable. This approach is often adopted in the machine learning community, placing a strong focus on optimizing a specific objective function to locate the optimal set of parameters.

More generally, we use the frequentist approach to find the correct answer about θ. For example, we can locate the value of θ by maximizing the joint probability of the actual data via *maximum likelihood estimation* (MLE), where the resulting solution is $\hat{\theta} = \mathrm{argmax}_\theta \, p(\mathrm{data}|\theta)$. There is no distribution involved with θ since we treat it as a fixed quantity, which makes the calculation easier as we only need to work with the probability distribution for the data. The final solution using the frequentist approach is a specific value of θ. And since we are working with samples that come from the underlying data-generating distribution, different samples would vary from each other, and the goal is to find the optimal parameter θ that best describes the current sample we are observing.

On the other hand, the Bayesian approach takes on the extra complexity by treating θ as a random variable with its own probability distribution, which gets updated as more data is collected. This approach offers a holistic view on all possible values of θ and the corresponding probabilities instead of the most probable value of θ alone. This is a different approach because the data is now treated as fixed and the parameter θ as a random variable. The optimal probability distribution for θ is then derived, given the observed fixed sample. There is no right or wrong in the Bayesian approach, only probabilities. The final solution is thus a probability distribution of θ instead of one specific value. Figure 1-9 summarizes these two different schools of thought.

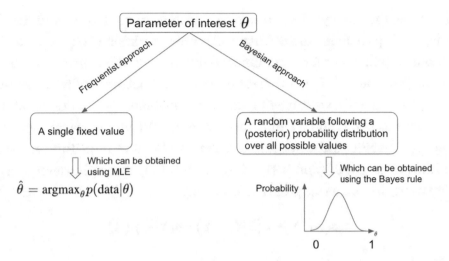

Figure 1-9. *Comparing the frequentist approach and the Bayesian approach regarding the parameter of interest. The frequentist approach treats θ as a fixed quantity that can be estimated via MLE, while the Bayesian approach employs a probability distribution which gets refreshed as more data is collected*

Joint, Conditional, and Marginal Probabilities

We have been characterizing the random variable θ using a (continuous) probability distribution $p(\theta)$. A probability distribution is a function that maps a specific value of θ to a probability, and the probabilities of all values of θ sum to one, that is, $\int p(\theta)d\theta = 1$.

Things become more interesting when we work with multiple (more than one) variables. Suppose we have two random variables x and y, and we are interested in two events $x = X$ and $y = Y$, where both X and Y are specific values that x and y may assume, respectively. Also, we assume the two random variables are dependent in some way. This would lead us to three types of probabilities commonly used in modern machine learning and Bayesian optimization literature: joint probability, marginal probability, and conditional probability, which we will look at now in more detail.

The joint probability of the two events refers to the probability of them occurring simultaneously. It is also referred to as the joint probability distribution since the probability now represents all possible combinations of the two simultaneous events.

We can write the joint probability of the two events as $p(X \text{ and } Y) = p(x = X \cap y = Y) = p(X \cap Y)$. Using the *chain rule* of probability, we can further write $p(X \text{ and } Y) = p(X \text{ given } Y) * p(Y) = p(X|Y)p(Y)$, where $p(X|Y)$ denotes the probability of event $x = X$

occurs given that the event $y = Y$ has occurred. It is thus referred to as conditional probability, as the probability of the first event is now conditioned on the second event. All conditional probabilities for a (continuous) random variable x given a specific value of another random variable (i.e., $y = Y$) form the conditional probability distribution $p(x|y = Y)$. More generally, we can write the joint probability distribution of random variables x and y as $p(x, y)$ and conditional probability distribution as $p(x \mid y)$.

The joint probability is also symmetrical, that is, $p(X \text{ and } Y) = p(Y \text{ and } X)$, which is a result of the *exchangeability* property of probability. Plugging in the definition of joint probability using the chain rule gives the following:

$$p(X \cap Y) = p(X|Y)p(Y) = p(Y|X)p(X)$$

If you look at this equation more closely, it is not difficult to see that it can lead to the Bayesian formula we introduced earlier, namely:

$$p(X|Y) = \frac{p(Y|X)p(X)}{p(Y)}$$

Understanding this connection gives us one more reason not to memorize the Bayesian formula but to appreciate it. We can also replace a single event $x = X$ with the random variable x to get the corresponding conditional probability distribution $p(x|y = Y)$.

Lastly, we may only be interested in the probability of an event for one random variable alone, disregarding the possible realizations of the other random variable. That is, we would like to consider the probability of the event $x = X$ under all possible values of y. This is called the marginal probability for the event $x = X$. The marginal probability distribution for a (continuous) random variable x in the presence of another (continuous) random variable y can be calculated as follows:

$$p(x) = \int p(x,y)dy = \int p(x|y)p(y)dy$$

The preceding definition essentially sums up possible values $p(x|y)$ weighted by the likelihood of occurrence $p(y)$. The weighted sum operation resolves the uncertainty in the random variable y and thus in a way integrates it out of the original joint probability distribution, keeping only one random variable. For example, the prior probability $p(\theta)$ in Bayes' rule is a marginal probability distribution of θ, which integrates out other random variables, if any. The same goes for the evidence term $p(\text{data})$ which is calculated by integrating over all possible values of θ.

Similarly, we have the marginal probability distribution for random variable y defined as follows:

$$p(y) = \int p(x,y)dx = \int p(y|x)p(x)dx$$

Figure 1-10 summarizes the three common probability distributions. Note that the joint probability distribution focuses on two or more random variables, while both the conditional and marginal probability distributions generally refer to a single random variable. In the case of the conditional probability distribution, the other random variable assumes a specific value and thus, in a way, "disappears" from the joint distribution. In the case of the marginal probability distribution, the other random variable is instead integrated out of the joint distribution.

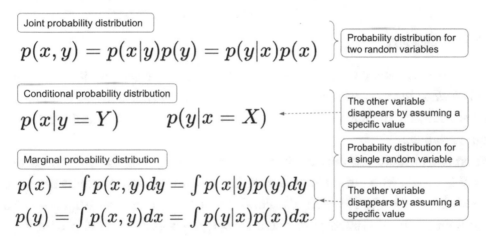

Figure 1-10.   *Three common probability distributions. The joint probability distribution represents the probability distribution for two or more random variables, while the conditional and marginal probability distributions generally refer to the probability distribution for one random variable. The conditional distribution represents the probabilities of a random variable by assuming/ conditioning a specific value for other variables, while the marginal distribution converts a joint probability to a single random variable by integrating out other variables*

Let us revisit Bayes' rule in the context of conditional and marginal probabilities. Specifically, the likelihood term $p(\text{data}|\theta)$ can be treated as the conditional probability of the data given the parameter θ, and the evidence term $p(\text{data})$ is a marginal probability that needs to be evaluated across all possible choices of θ. Based on the definition of marginal probability, we can write the calculation of $p(\text{data})$ as a weighted sum (assuming a continuous θ):

$$p(\text{data}) = \int p(\text{data}|\theta)p(\theta)d\theta$$

where we have a different likelihood conditioned by a specific parameter θ, and these likelihood terms are weighted by the prior probabilities. Thus, the evidence considers all the different ways we could use to get to the particular data.

Independence

A special case that would impact the calculation of the three probabilities mentioned earlier is *independence*, where the random variables are now independent of each other. Let us look at the joint, conditional, and marginal probabilities with independent random variables.

When two random variables are independent of each other, the event $x = X$ would have nothing to do with the event $y = Y$, that is, the conditional probability for $x = X$ given $y = Y$ becomes $p(X|Y) = p(X)$. The conditional probability distribution for two independent random variables thus becomes $p(x|y) = p(x)$. Their joint probability becomes the multiplication of individual probabilities: $p(X \cap Y) = P(X|Y)P(Y) = p(X)p(Y)$, and the joint probability distribution becomes a product of individual probability distributions: $p(x, y) = p(x)p(y)$. The marginal probability of x is just its own probability distribution:

$$p(x) = \int p(x|y)p(y)dy = \int p(x)p(y)dy = p(x)\int p(y)dy = p(x)$$

where we have used the fact that $p(x)$ can be moved out of the integration operation due to its independence with y, and the total area under a probability distribution is one, that is, $\int p(y)dy = 1$.

We can also extend to *conditional independence*, where the random variable x could be independent from y given another random variable z. In other words, we have $p(x, y|z) = p(x|z)p(y|z)$.

Prior and Posterior Predictive Distributions

Let us shift gear to focus on the actual predictions by quantifying the uncertainties using Bayes' rule. To facilitate the discussion, we will use y to denote the data in Bayes' formula or the actual observations as in the Bayesian optimization setting. We are interested in its predictive distribution $p(y)$, that is, the possible values of y and the corresponding probabilities. Our decision-making would be much more informed if we had a good understanding of the predictive distribution of the future unknown data, particularly in the Bayesian optimization framework where one needs to decide the next sampling location carefully.

Before we collect any data, we would work with a prior predictive distribution that considers all possible values of the underlying parameter θ. That is, the prior predictive distribution for y is a marginal probability distribution that could be calculated by integrating out all dependencies on the parameter θ:

$$p(y) = \int p(y, \theta) d\theta = \int p(y|\theta) p(\theta) d\theta$$

which is the exact definition of the evidence term in Bayes' formula. In a discrete world, we would take the prior probability for a specific value of the parameter θ, multiply the likelihood of the resulting data given the current θ, and sum across all weighted likelihoods.

Now let us look at the posterior predictive distribution for a new data point y' after observing a collection of data points collectively denoted as $\mathcal{D}$. We would like to assess how the future data would be distributed and what value of y' we would likely to observe if we were to run the experiment and acquire another data point again, given that we have observed some actual data. That is, we want to calculate the posterior predictive distribution $p(y'|\mathcal{D})$.

We can calculate the posterior predictive distribution by treating it as a marginal distribution (conditioned on the collected dataset $\mathcal{D}$) and applying the same technique as before, namely:

$$p(y'|\mathcal{D}) = \int p(y', \theta|\mathcal{D}) d\theta = \int p(y'|\theta, \mathcal{D}) p(\theta|\mathcal{D}) d\theta$$

where the second term $p(\theta|\mathcal{D})$ is the posterior distribution of the parameter θ that can be calculated by applying Bayes' rule. However, the first term $p(y'|\theta, \mathcal{D})$ is more involved. When assessing a new data point after observing some existing data points, a

19

common assumption is that they are *conditionally independent* given a particular value of θ. Such conditional independence implies that $p(y'|\theta,\mathcal{D})=p(y'|\theta)$, which happens to be the likelihood term. Thus, we can simplify the posterior predictive distribution as follows:

$$p(y'|\mathcal{D})=\int p(y'|\theta)p(\theta|\mathcal{D})d\theta$$

which follows the same pattern of calculation compared to the prior predictive distribution. This would then give us the distribution of observations we would expect for a new experiment (such as probing the environment in the Bayesian optimization setting) given a set of previously collected observations. The prior and posterior predictive distributions are summarized in Figure 1-11.

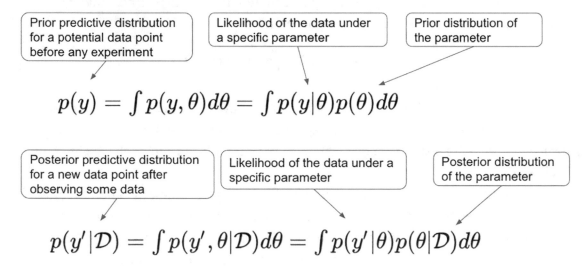

Figure 1-11. *Definition of the prior and posterior predictive distributions. Both are calculated based on the same pattern of a weighted sum between the prior and the likelihood*

Let us look at an example of the prior predictive distribution under a normal prior and likelihood function. Before the experiment starts, we assume the observation model for the likelihood of the data y to follow a normal distribution, that is, $y \sim N(\theta, \sigma^2)$, or $p(y|\theta, \sigma^2) = N(\theta, \sigma^2)$, where θ is the underlying parameter and σ^2 is a fixed variance. For example, in the case of the observation model in the Bayesian optimization setting introduced earlier, the parameter θ could represent the true objective function, and the variance σ^2 originates from an additive Gaussian noise. The distribution of y is dependent on θ, which itself is an uncertain quantity. We further assume the parameter θ to follow a normal distribution as its prior, that is, $\theta \sim N\left(\theta_0, \sigma_\theta^2\right)$, or $p(\theta) = N\left(\theta_0, \sigma_\theta^2\right)$, where θ_0 and σ_θ^2 are the mean and variance of our prior normal distribution assumed before collecting any data points. Since we have no knowledge of the environment of interest, we would like to understand how the data point (treated as a random variable) y could be distributed in this unknown environment under different values of θ.

Understanding the distribution of y upon the start of any experiment amounts to calculating its prior predictive distribution $p(y)$. Since we are working with a continuous θ, the marginalization needs to consider all possible values of θ from negative to positive infinity in order to integrate out the uncertainty due to θ:

$$p(y) = \int_{-\infty}^{\infty} p(y, \theta)\, d\theta = \int_{-\infty}^{\infty} p(y|\theta)\, p(\theta)\, d\theta$$

The prior predictive distribution can thus be calculated by plugging in the definition of normal likelihood term $p(y|\theta)$ and the normal prior term $p(\theta)$. However, there is a simple trick we can use to avoid the math, which would otherwise be pretty heavy if we were to plug in the formula of the normal distribution directly.

Let us try directly working with the random variables. We will start by noting that $y = (y - \theta) + \theta$. The first term $y - \theta$ takes θ away from y, which decentralizes y by changing its mean to zero and removes the dependence of y on θ. In other words, $(y - \theta) \sim N(0, \sigma^2)$, which also represents the distribution of the random noise in the observation model of Bayesian optimization. Since the second term θ is also normally distributed, we can derive the distribution of y as follows:

$$y \sim N\left(0, \sigma^2\right) + N\left(\theta_0, \sigma_\theta^2\right) = N\left(\theta_0, \sigma^2 + \sigma_\theta^2\right)$$

where we have used the fact that the addition of two independent normally distributed random variables will also be normally distributed, with the mean and variance calculated based on the sum of individual means and variances.

Therefore, the marginal probability distribution of y becomes $p(y) = N\left(\theta_0, \sigma^2 + \sigma_\theta^2\right)$. Intuitively, this form also makes sense. Before we start to collect any observation about y, our best guess for its mean would be θ_0, the expected value of the underlying random variable θ. Its variance is the sum of individual variances since we are considering uncertainties due to both the prior and the likelihood; the marginal distribution needs to absorb both variances, thus compounding the resulting uncertainty. Figure 1-12 summarizes the derivation of the prior predictive distributions under the normality assumption for the likelihood and the prior for a continuous θ.

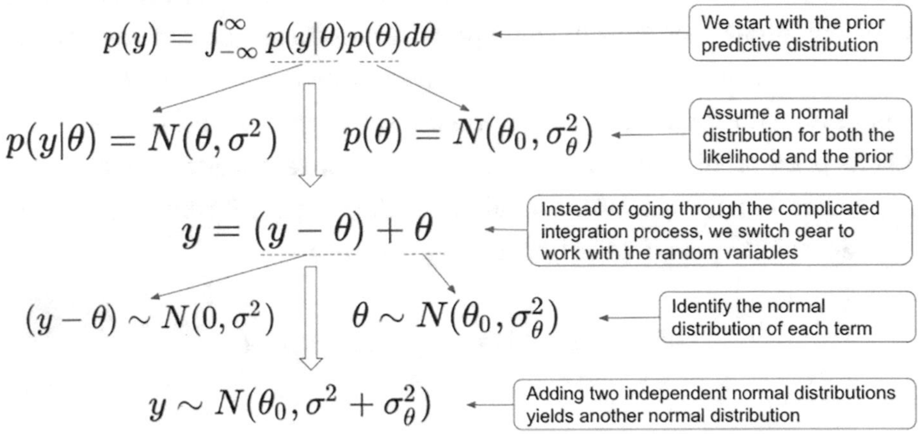

$$p(y) = \int_{-\infty}^{\infty} p(y|\theta)p(\theta)d\theta$$

We start with the prior predictive distribution

$$p(y|\theta) = N(\theta, \sigma^2) \quad \quad p(\theta) = N(\theta_0, \sigma_\theta^2)$$

Assume a normal distribution for both the likelihood and the prior

$$y = (y - \theta) + \theta$$

Instead of going through the complicated integration process, we switch gear to work with the random variables

$$(y - \theta) \sim N(0, \sigma^2) \quad \quad \theta \sim N(\theta_0, \sigma_\theta^2)$$

Identify the normal distribution of each term

$$y \sim N(\theta_0, \sigma^2 + \sigma_\theta^2)$$

Adding two independent normal distributions yields another normal distribution

Figure 1-12. *Derivation process of the prior predictive distribution for a new data point before collecting any observations, assuming a normal distribution for both the likelihood and the prior*

We can follow the same line of reasoning for the case of posterior predictive distribution for a new observation y' after collecting some data points $\mathcal{D}$ under the normality assumption for the likelihood $p(y'|\theta)$ and the posterior $p(\theta|\mathcal{D})$, where $p(y'|\theta) = N(\theta, \sigma^2)$ and $p(\theta|\mathcal{D}) = N(\theta', \sigma_\theta'^2)$. We can see that the posterior distribution for θ has an updated set of parameters θ' and $\sigma_\theta'^2$ using Bayes' rule as more data is collected.

Now recall the definition of the posterior predictive distribution with a continuous underlying parameter θ:

$$p(y'|\mathcal{D}) = \int_{-\infty}^{\infty} p(y'|\theta)p(\theta|\mathcal{D})d\theta$$

Again, plugging in the expression of two normally distributed density functions and working with an integration operation would be too tedious. We can instead write $y' = (y' - \theta) + \theta$, where $(y' - \theta) \sim N(0, \sigma^2)$ and $\theta \sim N(\theta', \sigma_\theta'^2)$. Adding the two independent normal distributions gives the following:

$$y' \sim N(0, \sigma^2) + N(\theta', \sigma_\theta'^2) = N(\theta', \sigma^2 + \sigma_\theta'^2)$$

Figure 1-13 summarizes the derivation of the posterior predictive distributions under normality assumption for the likelihood and the prior for a continuous θ.

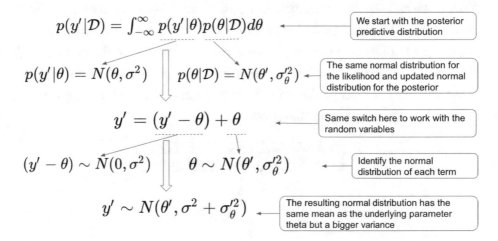

Figure 1-13. *Derivation process of the posterior predictive distribution for a new data point after collecting some observations, assuming a normal distribution for both the likelihood and the prior*

Bayesian Inference: An Example

After going through a quick and essential primer on Bayesian inference, let us put the mathematics in perspective by going through a concrete example. To start with, we will choose a probability density function for the prior $p(\theta) = N\left(\theta_0, \sigma_\theta^2\right)$ and the likelihood $p(y|\theta) = N(\theta, \sigma^2)$, both of which are normally distributed. Note that the prior and the likelihood are with respect to the random variables θ and y, respectively, where θ represents the true underlying objective value that is unknown, and y is the noise-corrupted actual observation that follows an observation model $y = \theta + \varepsilon$ with an additive and normally distributed random noise $\varepsilon \sim N(0, \sigma^2)$. We would like to infer the distribution of θ based on actual observed realization of y.

The choice for the prior is based on our subjective experience with the parameter of interest θ. Using the Bayes' theorem, it helps jump-start the learning toward its real probability distribution. The range of possible values of θ spans across the full support of the prior normal distribution. Upon observing an actual realization Y of y, two things will happen: the probability of observing $\theta = Y$ will be calculated and plugged in Bayes' rule as the prior, and the likelihood function will be instantiated as a conditional normal distribution $p(y|\theta = Y) = N(Y, \sigma^2)$.

Figure 1-14 illustrates an example of the marginal prior distribution and the conditional likelihood function (which is also a probability distribution) along with the observation Y. We can see that both distributions follow a normal curve, and the mean of the latter is aligned to the actual observation Y due to the conditioning effect from $Y = \theta$. Also, the probability of observing Y is not very high based on the prior distribution $p(\theta)$, which suggests a change needed for the prior in the posterior update of the next iteration. We will need to change the prior in order to improve such probability and conform the subjective expectation to reality.

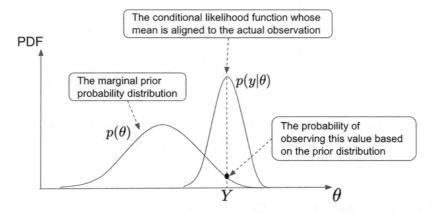

Figure 1-14. *Illustrating the prior distribution and the likelihood function, both following a normal distribution. The mean of the likelihood function is equal to the actual observation due to the effect of conditioning*

The prior distribution will then gradually get updated to approximate the actual observations by invoking Bayes' rule. This will give the posterior distribution $p(\theta|Y) = N(\theta', \sigma_\theta'^2)$ in solid line, whose mean is slightly nudged from θ_0 toward Y and updated to θ', as shown in Figure 1-15. The prior distribution and likelihood function are displayed in dashed lines for reference. The posterior distribution of θ is now more aligned with what is actually observed in reality.

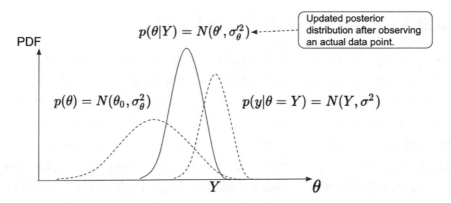

Figure 1-15. *Deriving the posterior distribution for θ using Bayes' rule. The updated mean θ′ is now between the prior mean $θ_0$ and actual observation Y, suggesting an alignment between subjective preference and reality*

Finally, we would be interested in the predictive distribution of the actual data point if we acquired a new observation. Treating it as a random variable y' enables us to express our uncertainty in the form of an informed probability distribution, which benefits follow-up tasks such as deciding where to sample next (more on this later). Based on our previous discussion, the resulting probability distribution for y' will assume a normal distribution with the same mean as the posterior distribution of θ and an inflated variance that absorbs uncertainties from both θ and the observation model for y, as shown in Figure 1-16. The prior and posterior distributions and the likelihood function are now in the dashed line for reference.

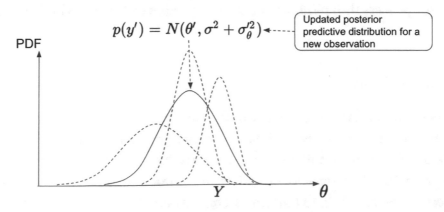

Figure 1-16. *Illustrating the posterior predictive distribution if we acquire another observation from the system/environment under study. The posterior predictive distribution shares the same mean as the posterior distribution of θ but now has a larger spread due to uncertainty from both θ and the observation model*

Bayesian Optimization Workflow

Having gone through the essentials in Bayesian statistics, you may wonder how it connects to the Bayesian optimization setting. Recall that the predictive posterior distribution quantifies the probabilities of different observations if we were to probe the environment and receive a new realization. This is a powerful tool when reasoning about the utility of varying sampling choices $x \in A$ in search of the global maximum f^*.

To put this in perspective, let us add the location x as an explicit conditioning on the prior/posterior predictive distribution and use f (the true objective value) to denote θ (the underlying parameter of interest). For example, the prior predictive distribution $p(y|x)$ represents the conditional probability distribution of the actual observation y at location x. Since we have many different locations across the domain A, there will also be many prior predictive distributions, each following the same class of probability distributions when assuming specific properties about the underlying true objective function before probing the environment.

Take a one-dimensional objective function $f(x) \in R$, for example. For any location $x_0 \in A$, we will use a prior predictive distribution $p(y|x_0)$ to characterize the possible values of the unknown true value $f_0 = f(x_0)$. These prior predictive distributions will jointly form our prior probabilistic belief about the shape of the underlying objective function f. Since there are infinitely many locations x_0 and thus infinitely many random variables y, these probability distributions of the infinite collection of random variables are jointly used as a *stochastic process* to characterize the true objective function. Here, the stochastic process simplifies to our running example earlier when limited to a single location.

Gaussian Process

A prevalent choice of stochastic process in Bayesian optimization is the *Gaussian process*, which requires that these finite-dimensional probability distributions are multivariate Gaussian distributions in a continuous domain with infinite number of variables. It is a flexible framework to model a broad family of functions and quantify their uncertainties, thus being a powerful *surrogate model* used to approximate the true underlying function. We will delve into the details of the Gaussian process in the next chapter, but for now, let us look at a few visual examples to see what it offers.

Figure 1-17 illustrates an example of a "flipped" prior probability distribution for a single random variable selected from the prior belief of the Gaussian process. Each point follows a normal distribution. Plotting the mean (solid line) and 95% credible interval (dashed lines) of all these prior distributions gives us the prior process for the objective function regarding each location in the domain. The Gaussian process thus employs an infinite number of normally distributed random variables within a bounded range to model the underlying objective function and quantify the associated uncertainty via a probabilistic approach.

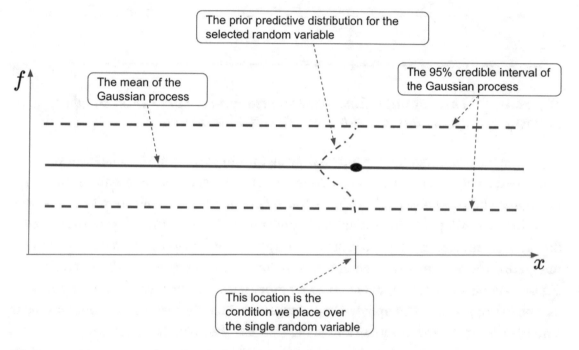

Figure 1-17. *A sample prior belief of the Gaussian process represented by the mean and 95% credible interval for each location in the domain. Every objective value is modeled by a random variable that follows a normal prior predictive distribution. Collecting the distributions of all random variables could help us quantify the potential shape of the true underlying function and its probability*

The prior process can thus serve as the surrogate data-generating process to generate samples in the form of functions, an extension of sampling single points from a probability distribution. For example, if we were to repeatedly sample from the prior process earlier, we would expect the majority (around 95%) of the samples to fall within the credible interval and a minority outside this range. Figure 1-18 illustrates three functions sampled from the prior process.

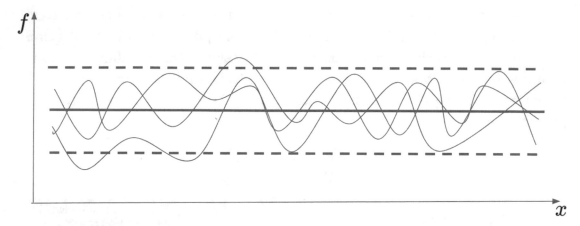

Figure 1-18. *Three example functions sampled from the prior process, where majority of the functions fall within the 95% credible interval*

In the Gaussian process, the uncertainty on the objective value of each location is quantified using the credible interval. As we start to collect observations and assume a noise-free and exact observation model, the uncertainties at the collection locations will be resolved, leading to zero variance and direct interpolation at these locations. Besides, the variance increases as we move further away from the observations, resulting from integrating the prior process with the information provided by the actual observations. Figure 1-19 illustrates the updated posterior process after collecting two observations. The posterior process with updated knowledge based on the observations will thus make a more accurate surrogate model and better estimate the objective function.

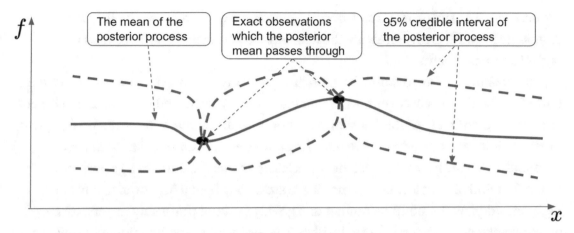

Figure 1-19. *Updated posterior process after incorporating two exact observations in the Gaussian process. The posterior mean interpolates through the observations, and the associated variance reduces as we move nearer the observations*

Acquisition Function

The tools from Bayesian inference and the extension to the Gaussian process provide principled reasoning on the distribution of the objective function. However, we would still need to incorporate such probabilistic information in our decision-making to search for the global maximum. We need to build a policy that absorbs the most updated information on the objective function and recommends the following most promising sampling location in the face of uncertainties across the domain. The optimization policy thus plays an essential role in connecting the Gaussian process to the eventual goal of Bayesian optimization. In particular, the posterior predictive distribution provides an outlook on the objective value and associated uncertainty for locations not explored yet, which could be used by the optimization policy to quantify the *utility* of any alternative location within the domain.

When converting the posterior knowledge about candidate locations, that is, posterior parameters such as the mean and the variance, to a single utility score, the acquisition function comes into play. An acquisition function is a manually designed mechanism that evaluates the relative potential of each candidate location in the form of a scalar score, and the location with the maximum score will be used as the recommendation for the next round of sampling. It is a function that assesses how valuable a candidate location when we acquire/sample it. The acquisition function

29

is often cheap to evaluate as a side computation since we need to evaluate it at every candidate location and then locate the maximum utility score, another (inner) optimization problem.

Many choices of acquisition function have been proposed in the literature. In a later part of the book, we will cover the popular ones, such as expected improvement (EI) and knowledge gradient (KG). Still, it suffices, for now, to understand that it is a predesigned function that needs to balance two opposing forces: *exploration* and *exploitation*. Exploration encourages resolving the uncertainty across the domain by sampling at unfamiliar and distant locations, since these areas may bear a big surprise due to a high certainty. Exploitation recommends a greedy move at promising regions where we expect the observation value to be high. The exploration-exploitation trade-off is a common topic in many optimization settings.

Another distinguishing feature is the short-term and long-term trade-off. A short-term acquisition function only focuses on one step ahead and assumes this is the last chance to sample from the environment; thus, the recommendation is to maximize the immediate utility. A long-term acquisition function employs a multi-step lookahead approach by simulating potential evolutions/paths in the future and making a final recommendation by maximizing the long-run utility. We will cover both types of policies in the book.

There are many other emerging variations in the design of the acquisition function, such as adding safety constraints to the system under study. In any case, we would judge the quality of the policy using a specific acquisition function based on how close we are to the location of the global maximum upon exhausting our budget. The distance between the current and optimal locations is often called *instant regret or simple regret*. Alternatively, the *cumulative regret* (cumulative distances between historical locations and the optimum location) incurred throughout the sampling process can also be used.

The Full Bayesian Optimization Loop

Bayesian optimization is an iterative process between the (uncontrolled) environment and the (controlled) policy. The policy involves two components to support sequential decision-making: a Gaussian process as the surrogate model to approximate the true underlying function (i.e., the environment) and an acquisition function to recommend the following best sampling location. The environment receives the probing request at a specific location and responds by revealing a new observation that follows a particular observation model. The Gaussian process surrogate model then uses the new observation to obtain a posterior process in support of follow-up decision-making by the preset acquisition function. This process continues until the stopping criterion such as exhausting a given budget is met. Figure 1-20 illustrates this process.

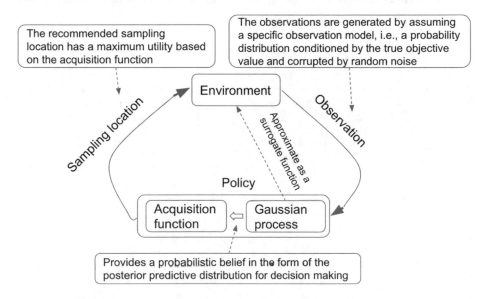

Figure 1-20. *The full Bayesian optimization loop featuring an iterative interaction between the unknown (black-box) environment and the decision-making policy that consists of a Gaussian process for probabilistic evaluation and acquisition function for utility assessment of candidate locations in the environment*

Summary

Bayesian optimization is a class of methodology that aims at sample-efficient global optimization. This chapter covered the foundations of the BO framework, including the following:

- The defining characteristics of Bayesian optimization as a global optimization technique, including the observation model, surrogate model, and the acquisition function

- The basics of Bayesian statistics, including the Bayesian inference framework, different types of probabilities (joint, conditional, and marginal), and prior and posterior predictive distributions

- The Bayesian optimization workflow that highlights two major components: the Gaussian process and the acquisition function

In the next chapter, we will discuss the first component: the Gaussian process, covering both theoretical understanding and practical implementation in Python.

CHAPTER 2

Gaussian Processes

In the previous chapter, we covered the derivation of the posterior distribution for parameter θ as well as the predictive posterior distribution of a new observation y' under a normal/Gaussian prior distribution. Knowing the posterior predictive distribution is helpful in supervised learning tasks such as regression and classification. In particular, the posterior predictive distribution quantifies the possible realizations and uncertainties of both existing and future observations (if we were to sample again). In this chapter, we will cover some more foundation on the Gaussian process in the first section and switch to the implementation in code in the second section.

The way we work with the parameters depends on the type of models used for training. There are two types of models in supervised learning tasks: *parametric* and *nonparametric* models. Parametric models assume a fixed set of parameters to be estimated and used for prediction. For example, by defining a set of parameters $\boldsymbol{\theta}$ (bolded lowercase to denote multiple elements contained in a vector) given a set of input observations $\mathbf{X}$ (bolded uppercase to denote a matrix) and output target $\mathbf{y}$, we rely on the parametric model $p(\mathbf{y}|\mathbf{X}, \boldsymbol{\theta})$ and estimate the optimal parameter values $\hat{\theta}$ via procedures such as maximum likelihood estimation or maximum a posteriori estimation. Using a Bayesian approach, we can also infer the full posterior distribution $p(\boldsymbol{\theta}|\mathbf{X}, \mathbf{y})$ to enable a distributional representation instead of a point estimate for the parameters $\boldsymbol{\theta}$.

Figure 2-1 illustrates the shorthand math notation for matrix $\mathbf{X}$ and vector $\mathbf{y}$.

P. Liu, *Bayesian Optimization*, https://doi.org/10.1007/978-1-4842-9063-7_2

Input matrix

Output vector

$$\mathbf{X} = \begin{bmatrix} \mathbf{x}_1^\top \\ \mathbf{x}_2^\top \\ \vdots \\ \mathbf{x}_n^\top \end{bmatrix} = \begin{bmatrix} x_1^1 & x_1^2 & \cdots & x_1^p \\ x_2^1 & x_2^2 & \cdots & x_2^p \\ \vdots & \vdots & \ddots & \vdots \\ x_n^1 & x_n^2 & \cdots & x_n^p \end{bmatrix} \qquad \mathbf{y} = \begin{bmatrix} y_1 \\ y_2 \\ \vdots \\ y_n \end{bmatrix}$$

Figure 2-1. Visualizing the input matrix and output vector

On the other hand, nonparametric methods do not assume a fixed set of parameters, and, instead, the number of parameters is determined by the size of the dataset. For example, the *k-nearest neighbors* (KNN) algorithm is a nonparametric model that takes the average of the nearest k data points as the final prediction. This process involves calculating the distance/similarity of all data points, where these similarities serve as the weights (equal weightage for all k nearby points in KNN). Therefore, it tends to be slow at test time when the size of the dataset is large, although training time is faster compared with parametric models due to direct data memorization.

Gaussian processes (GP) can be considered a type of nonparametric model. It is a stochastic process used to characterize the *distribution over functions* instead of a fixed set of parameters. The key difference is that GP extends a limited set of parameters θ from a discrete space, often used in multivariate Gaussian distribution, into an unlimited function *f* in a continuous and infinite space, which corresponds to a function or a curve when plotted on a graph. GP inherits the nice mathematical properties of a multivariate Gaussian distribution and offers a flexible framework to model functions with uncertainty estimates.

For example, Figure 2-2 shows one realization of a coefficient vector θ that contains three parameters and a function *f* that displays as a curve and has infinite number of parameters. However, computationally we would still use a finite number of points to approximate the actual curve when plotting the function. Using GP, the function *f* represents an infinite collection of random variables, where any (finite) subset of these random variables follows a joint Gaussian distribution. The mutual dependence among these random variables thus determines the resulting shape of the function f.

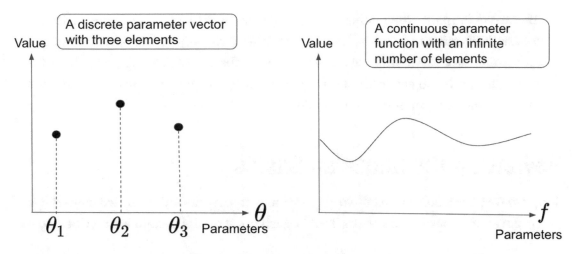

Figure 2-2. *Comparing a discrete parameter vector (left) and a continuous parameter function (right), which is an extension from a limited set of parameters to an unlimited number of parameters embedded in a function*

Since an arbitrary multivariate Gaussian distribution can be characterized by its mean vector and positive semidefinite covariance matrix, we can extend the same to functions that span across a continuous domain A and characterize a GP via a *mean function* $\mu : A \rightarrow \mathbb{R}$ and a positive semidefinite *covariance function* $\kappa : A \times A \rightarrow \mathbb{R}$:

$$p(f) = GP(f;\mu,\kappa)$$

The mean function μ returns the expected value (central tendency) for an arbitrary input location x, and the covariance function quantifies the similarity between any two input locations.

We will shed more light on the covariance function later. For now, it helps to think of GP as a massive Gaussian distribution. Under the GP framework, inference on the distribution over functions starts with a prior process (also over the functions), which gets iteratively updated as new observations are revealed. By treating f as a collection of random variables that follow a GP prior and using Bayesian inference to obtain the GP posterior, we could quantify the uncertainties about the underlying true objective function itself, including its maximizer f^* and x^*. Also, depending on the type of task, GP can be used for both regression and classification.

Before we formalize the mathematical properties of GP, let us review the basics of Gaussian distribution in a multivariate setting, which will help us build the intuition in understanding GP. We will then delve into the mechanics of developing useful prior distributions for the objective function and calculating the posterior distributions after observing some noisy or noise-free samples.

Reviewing the Gaussian Basics

Suppose we have a collection of two-dimensional observations distributed centered at the origin over a coordinate system. That is, each point on the system is described by two scalar-valued features, that is, $\mathbf{x} = \begin{bmatrix} x^1 \\ x^2 \end{bmatrix} \in \mathbb{R}^2$. Here, $\mathbb{R}^2$ describes a two-dimensional plane that consists of all real numbers extended to infinity. These points could be equally distributed around the origin or with some systematic pattern, resulting in the two scenarios shown in Figure 2-3.

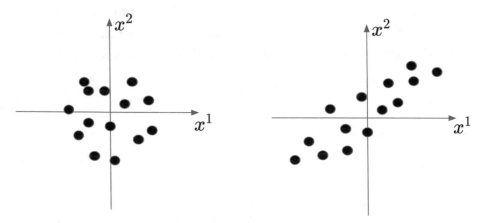

Figure 2-3. *Illustrating two possible distributions of points on a plane centered at the origin. These points could be equally distributed around the origin (left) or distributed with a systematic bias (right)*

Now we would like to fit a Gaussian distribution for each scenario, which will allow us to describe other unknown points better if the distribution is fit properly. In the two-dimensional case, a Gaussian distribution is characterized by a mean vector $\mu = \begin{bmatrix} \mu_1 \\ \mu_2 \end{bmatrix}$ and a covariance matrix $\mathbf{K} = \begin{bmatrix} \sigma_{11}^2 & \sigma_{12}^2 \\ \sigma_{21}^2 & \sigma_{22}^2 \end{bmatrix}$. The mean vector describes the central tendency if we were to sample from the Gaussian distribution repeatedly, and the covariance matrix the intercorrelation among the points.

Correspondingly, any point on the plane can be considered as a realization (or equivalently, a sample or simulation) from the Gaussian distribution:

$$\mathbf{x} \sim N(\mathbf{\mu}, \mathbf{K}) = N\left(\begin{bmatrix} \mu_1 \\ \mu_2 \end{bmatrix}, \begin{bmatrix} \sigma_{11}^2 & \sigma_{12}^2 \\ \sigma_{21}^2 & \sigma_{22}^2 \end{bmatrix} \right)$$

Since we assume these points are centered around the origin, this implies that $\mathbf{\mu} = \begin{bmatrix} 0 \\ 0 \end{bmatrix}$. Now let us take a closer look at the covariance matrix.

Understanding the Covariance Matrix

The multivariate covariance matrix describes how the features of the data are related to each other. In the previous example, the covariance matrix tells us how x^2 changes in direction and magnitude if x^1 increases in value, that is, the co-movement between the two variables. Each entry in the covariance matrix is calculated using the definition of variance and covariance. For example:

$$\sigma_{11}^2 = var\left(x^1\right) = \mathbb{E}\left[\left(x^1 - \mathbb{E}\left[x^1\right]\right)^2 \right] = \mathbb{E}\left[\left(x^1\right)^2 \right]$$

$$\sigma_{12}^2 = \sigma_{21}^2 = \mathbb{E}\left[\left(x^1 - \mathbb{E}\left[x^1\right]\right)\left(x^2 - \mathbb{E}\left[x^2\right]\right) \right] = \mathbb{E}\left[x^1 x^2 \right]$$

where we have used the fact that $\mathbb{E}\left[x^1\right] = \mathbb{E}\left[x^2\right] = 0$, which is a result of centering the data points. The entry σ_{12}^2 (or equivalently σ_{21}^2) measures the similarity among observations between the x^1 dimension and the x^2 dimension, calculated using the dot product $x^1 x^2$. The dot product is a widely used approach to gauge the similarity between two vectors in terms of the magnitude of similarity and the direction.

The covariance is closely related to the concept of correlation. For variables x^1 and x^2, their correlation $\rho_{x^1 x^2}$ is defined as follows:

$$\rho_{x^1 x^2} = \frac{cov\left(x^1, x^2\right)}{sd\left(x^1\right) sd\left(x^2\right)} = \frac{\sigma_{12}^2}{\sigma_{11}\sigma_{22}} = \frac{\sigma_{21}^2}{\sigma_{11}\sigma_{22}}$$

where we have used the symmetric property of the covariance matrix assumed to be positive semidefinite, that is, $\sigma_{12}^2 = \left(\sigma_{21}^2 \right)^T$. The correlation is a metric that summarizes the covariance matrix by making use of all the components in the calculation. It is a numeric number between -1 and 1, where 1 represents perfect positive correlation (such as $x^2 = 2x^1$) and -1 means perfect negative correlation (such as $x^2 = -2x^1$).

Since the diagonal entries in the covariance matrix denote the feature-specific variance, we can easily apply normalization by dividing the standard deviation and obtain a unit variance, resulting in $\sigma_{11}^2 = \sigma_{22}^2 = 1$.

Things become more interesting when we look at the off-diagonal entries, that is, σ_{12}^2 and σ_{21}^2. In the left distribution in Figure 2-3, if we were to increase x^2 from zero to a specific value (say $x^2 = 0.8$), our knowledge of the possible values of x^1 (i.e., positive or negative) is still very limited. In fact, knowing x^2 does not contribute to our knowledge about x^1 at all, which still remains equally distributed around the vertical axis. The information about x^2 does not help getting more information about x^1. In other words, x^1 and x^2 are uncorrelated to each other, and $\sigma_{12}^2 = \sigma_{21}^2 = 0$. Therefore, for the first scenario, we have

$$\begin{bmatrix} x^1 \\ x^2 \end{bmatrix} \sim N \left(\begin{bmatrix} 0 \\ 0 \end{bmatrix}, \begin{bmatrix} 1 & 0 \\ 0 & 1 \end{bmatrix} \right)$$

On the other hand, if we were to take the same cut in the second scenario and condition x^2 to the same value, we know that x^1 will be positive with a high probability. This time, knowing the value of x^2 gives us some information about x^1, thus making these two variables correlated. We would expect x^1 to increase if x^2 were to increase, thus forming a positive correlation. Assume the covariance between x^1 and x^2 is 0.6, that is, the strength of co-movement is $\sigma_{12}^2 = \sigma_{21}^2 = 0.6$, the *bivariate* Gaussian distribution for the second scenario could then be expressed as

$$\begin{bmatrix} x^1 \\ x^2 \end{bmatrix} \sim N \left(\begin{bmatrix} 0 \\ 0 \end{bmatrix}, \begin{bmatrix} 1 & 0.6 \\ 0.6 & 1 \end{bmatrix} \right)$$

Figure 2-4 summarizes these two scenarios, where the key difference is the dependence between these two variables. A value of zero at the off-diagonal entries indicates that the variables are not dependent upon each other, while a positive or negative value suggests some level of dependence.

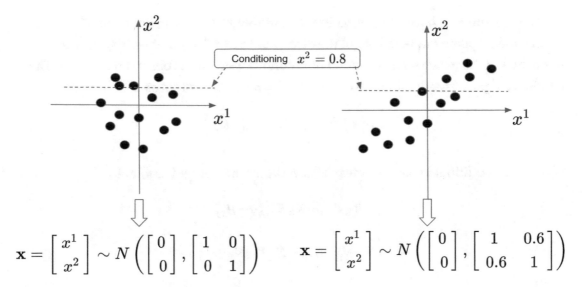

Figure 2-4. *Illustrating the bivariate Gaussian distribution for the two scenarios. The covariance entries on the left are set to zero to represent the fact that the variables are uncorrelated, while the covariance entries are 0.6 due to a positive correlation on the right*

Marginal and Conditional Distribution of Multivariate Gaussian

Our bivariate Gaussian example consists of two variables/features: x^1 and x^2, jointly Gaussian. When working with a multivariate Gaussian distribution, we are often interested in feature-wise distribution in terms of its marginal and conditional distribution. Having more insight into the individual distributions would allow us to systematically reason about the uncertainties of other unobserved variables in the Bayesian optimization process.

Recall that the bivariate Gaussian distribution is parameterized by a mean vector $\mu = \begin{bmatrix} \mu_1 \\ \mu_2 \end{bmatrix}$ and a covariance matrix $\mathbf{K} = \begin{bmatrix} K_{11} & K_{12} \\ K_{21} & K_{22} \end{bmatrix} = \begin{bmatrix} \sigma_{11}^2 & \sigma_{12}^2 \\ \sigma_{21}^2 & \sigma_{22}^2 \end{bmatrix}$. Based on the *multivariate Gaussian theorem*, we can write the marginal distribution for both variables as follows:

$$p(x^1) = N(x^1 | \mu_1, K_{11})$$

$$p(x^2) = N(x^2 | \mu_2, K_{22})$$

Now assume we have an observation for variable x^2, say $x^2 = a$, how does this information update our belief about the distribution of x^1? In other words, we are interested in the posterior distribution of x^1 conditioned on the observation $x^2 = a$. The conditional posterior distribution of x^1 given $x^2 = a$ can be written as

$$p\left(x^1|x^2=a\right)=N\left(x^1|\mu_{1|2},K_{1|2}\right)$$

where the conditional posterior mean and variance are defined as follows:

$$\mu_{1|2}=\mu_1+K_{12}K_{22}^{-1}\left(a-\mu_2\right)$$

$$K_{1|2}=K_{11}-K_{12}K_{22}^{-1}K_{21}$$

Note that the posterior variance of x^1 does not depend on the observed value of x^2.

Sampling from a Gaussian Distribution

In the previous sections, we covered how to infer the parameters of a multivariate Gaussian distribution from a collection of data points. By conditioning on these observations, we could obtain an updated view of the posterior distribution of other unobserved variables. We can then utilize the updated posterior distribution to sample and generate new imaginary observations for these unobserved variables to be revealed in the future. In this process, our posterior belief will get reinforced if the imagined samples do not deviate much from the actual observations.

Now assume we would like to sample from a Gaussian distribution $N(\mu, \sigma^2)$. How could we generate random samples that follow this specific distribution? One common approach is to first generate a random number x from a standard normal distribution $N(0, 1)$ and then apply the *scale-location transformation* to obtain a sample $\sigma x + \mu$. By scaling the sample based on the standard deviation σ followed by adding the mean μ, the resulting sample will follow a normal distribution with mean μ and variance σ^2.

The first step of sampling from a standard normal distribution is more involved, whereas the second step is a simple and deterministic transformation. A general approach is to transform a uniform random variable using the inverse cumulative distribution function of the standard Gaussian. For example, if U is uniformly distributed on $[0, 1]$, then $\Phi^{-1}(U)$ would follow a standard normal distribution, where Φ^{-1} is the inverse of the cumulative function of a standard normal distribution.

A visual example is shown in Figure 2-5. We start by sampling a random point from the uniform distribution on $[0, 1]$, followed by invoking the inverse cumulative density function (CDF) of the standard normal distribution to obtain the corresponding sample at the input space of CDF, given the fact that CDF monotonically maps an arbitrary input value to an output on $[0, 1]$. Mathematically, the random sample of the standard normal is given by $x = \Phi^{-1}(U)$. Considering the one-to-one mapping relationship between the probability density function (PDF) and the CDF of the standard normal, we could obtain the same input x at the PDF space as well. We would also expect most of the samples to be centered around the mean of the PDF. Finally, we apply the scale-location transformation to convert to the random sample of the normal distribution with the desired mean and variance.

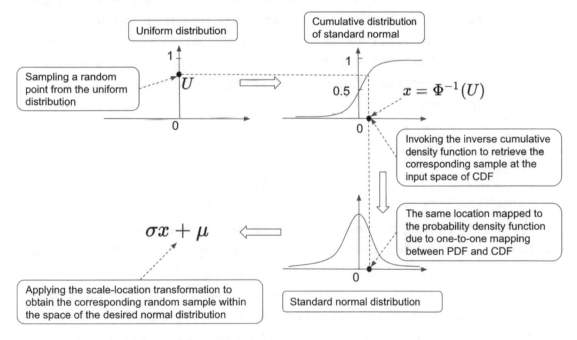

Figure 2-5. *Obtaining a random sample from the desired univariate Gaussian distribution based on three steps: sampling from a uniform distribution, converting to the corresponding input of the CDF (or PDF) using the inverse cumulative function, and applying the scale-location transformation to convert the random sample to the space of the normal distribution with the desired mean and variance*

Let us extend to the multivariate case and look at how to sample from a bivariate Gaussian distribution with an arbitrary mean vector $\mu = \begin{bmatrix} \mu_1 \\ \mu_2 \end{bmatrix}$ and covariance matrix $\mathbf{K} = \begin{bmatrix} K_{11} & K_{12} \\ K_{21} & K_{22} \end{bmatrix}$. We could follow a similar process and start by sampling from a standard bivariate normal distribution $N\left(\begin{bmatrix} 0 \\ 0 \end{bmatrix}, \begin{bmatrix} 1 & 0 \\ 0 & 1 \end{bmatrix} \right)$ and then apply the scale-location transformation to the resulting samples.

How could we randomly sample $\mathbf{x} = \begin{bmatrix} x^1 \\ x^2 \end{bmatrix}$ from $N\left(\begin{bmatrix} 0 \\ 0 \end{bmatrix}, \begin{bmatrix} 1 & 0 \\ 0 & 1 \end{bmatrix} \right)$? Put equivalently, we would like to sample $\mathbf{x}$ from $N(\mathbf{0}, \mathbf{I})$, where $\mathbf{0}$ is a vector of zeros and $\mathbf{I}$ is the identity matrix. Since the off-diagonal entries of the covariance matrix are zero, we could exploit the fact that the two variables x^1 and x^2 are uncorrelated. Thus, the problem becomes drawing random samples from the individual distribution for x^1 and x^2, respectively. Based on the previous theorem on multivariate Gaussian distribution, the marginal distributions are the univariate Gaussian distributions with the respective mean and variance. In other words:

$$x^1 \sim N(0,1)$$

$$x^2 \sim N(0,1)$$

Since we know already how to sample from a univariate standard normal distribution, we could sample a vector of two random samples for x^1 and x^2 and take the vector as a random sample from the bivariate standard normal distribution.

Next, we need to access the square root of the covariance matrix $\mathbf{K}$ in order to apply the scale-location transformation. One way to achieve this is to apply the *Cholesky decomposition*, which could decompose the matrix $\mathbf{K}$ into the product of a lower triangular matrix $\mathbf{L}$ and its transpose $\mathbf{L}^T$:

$$\mathbf{K} = \mathbf{L}\mathbf{L}^T$$

Thus, we can obtain a transformed sample via $\mathbf{L}\mathbf{x} + \mu$, which will follow $N(\mu, \mathbf{K})$.

Understanding the sampling process of a multivariate Gaussian distribution is essential when it comes to simulating samples (specifically, curves) from a posterior Gaussian process in the Bayesian framework. A random sample from the multivariate Gaussian distribution lives in the form of a vector with limited elements, which gets stretched to an unlimited function under Gaussian processes. Next, we will move into the setting of sequentially obtaining new observations and look at how to plug in the GP framework to obtain updated samples using GP regression.

Gaussian Process Regression

A Gaussian process is a tool for us to perform surrogate modeling by placing a prior belief over random functions and updating the posterior over these functions by sequentially observing new data points. Given a set of input-output pairs as observations, GP can be used in these regression settings to perform prediction that comes with uncertainty quantification. It is also referred to as *krigging* in geostatistics. Based on a GP governed by a specific prior covariance function, we could obtain an updated posterior covariance function using the observations, similar to updating the model parameters using the training data in the field of machine learning. In fact, Gaussian process regression is a nonparametric Bayesian approach to regression, which is widely used in practice due to the uncertainty measurements provided on the predictions.

Choosing a good covariance function, or the *kernel* function, is crucial as it determines the similarity between different samples and impacts the shape of the resulting functional curve from the GP. To better characterize the effect of a kernel function, let us go through an example of a regression case.

The Kernel Function

A kernel or covariance function $\kappa(x_i, x_j)$ measures the statistical relationship between two points x_i and x_j in the input space. Specifically, it quantifies the correlation between the change in x_j and the corresponding change in x_i in the GP framework. It is a function that outputs a scalar similarity value for any two points in the input space, typically defined over the Euclidean distance.

Suppose we have three pairs of noise-free observations $D = \{(x_1, f_1), (x_2, f_2), (x_3, f_3)\}$ where $\{x_i, i = 1...3\}$ represent the input locations and $\{f_i = f(x_i)\}$ are the corresponding functional value modeled as random variables. We want to model these three variables using a multivariate Gaussian distribution with a mean vector $\boldsymbol{\mu}$ and a covariance matrix $\mathbf{K}$. Without loss of generality, it is often assumed that the mean vector $\boldsymbol{\mu}=\mathbf{0}$, meaning the variables are centered to simplify the derivation. Also, note that we use $\{x_i\}$ with the subscript i to denote the location of the observations arranged along the horizontal axis, which are considered given in the Bayesian inference framework in the previous chapter.

We can assume the three variables jointly follow a multivariate Gaussian distribution as follows:

$$\mathbf{f} = \begin{bmatrix} f_1 \\ f_2 \\ f_3 \end{bmatrix} \sim N(\mu, \mathbf{K}) = N\left(\begin{bmatrix} 0 \\ 0 \\ 0 \end{bmatrix}, \begin{bmatrix} K_{11} & K_{12} & K_{13} \\ K_{21} & K_{22} & K_{23} \\ K_{31} & K_{32} & K_{33} \end{bmatrix} \right)$$

where the covariance function $\mathbf{K}$ captures the correlation between the three observations. Specifically, f_1 should be more correlated to f_2 than f_3 due to a smaller difference in value. Such prior belief also constitutes our inductive bias into the model: the resulting model should be smooth without sudden jumps between nearby points. See Figure 2-6 for a recap of this example.

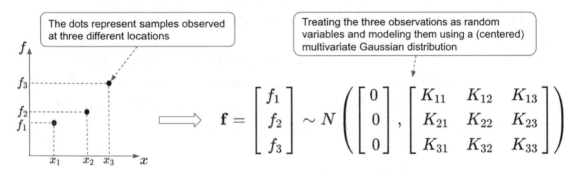

Figure 2-6. *Modeling the three samples using a multivariate Gaussian distribution*

Given the positive relationship between the similarity of observations (modeled as random variables) and their distance, we would expect K_{12} to be bigger than K_{13}. The same logic goes for other entries in the covariance matrix. Such insight could give rise to the following example covariance matrix after standardizing all data points (so that all diagonal entries are one):

$$\begin{bmatrix} f_1 \\ f_2 \\ f_3 \end{bmatrix} \sim N\left(\begin{bmatrix} 0 \\ 0 \\ 0 \end{bmatrix}, \begin{bmatrix} 1 & 0.8 & 0.3 \\ 0.8 & 1 & 0.7 \\ 0.3 & 0.7 & 1 \end{bmatrix} \right)$$

A popular kernel function we can use to obtain a covariance matrix similar to the preceding example is the squared exponential kernel:

$$K_{ij} = \kappa\left(x_i, x_j\right) = e^{-\|x_i - x_j\|^2}$$

where we take the exponential of the negative squared distance as the final distance measure. We can now quantify the similarity between any pair of points in the bounded range. For example, when x_i is very far away from x_j and their distance approaches infinity, that is, $\|x_i - x_j\| \to \infty$, we have $K_{ij} \to 0$; when $x_i = x_j$, that is, the similarity of a given point with itself, we have $K_{ij} = 1$, corresponding to the diagonal entries in the covariance matrix. Therefore, the squared exponential kernel function $\kappa(x_i, x_j)$ describes the similarity between any two input locations using a value between 0 and 1. Using such a kernel function would thus enable us to build a covariance matrix similar to the example earlier.

Let us implement the kernel function in the following code listing and observe the similarity value as the input distance increases.

Listing 2-1. Calculating the squared exponential distance

```
# define a kernel function to return a squared exponential distance
def kernel(x):
    return np.exp(-x**2)
# create a list of differences between two input locations
X = np.linspace(-5,5,100).reshape(-1,1)
K = kernel(X)
plt.plot(X, K)
```

Running the preceding code would generate Figure 2-7, which shows that the covariance between two points decreases as their distance increases and approaches zero when the distance exceeds two. This means that the functional values at nearby locations are more correlated, while distant locations would essentially lead to independent random variables with little or no covariance. It also suggests that revealing an observation at an arbitrary input location provides nontrivial information about the functional values of nearby locations.

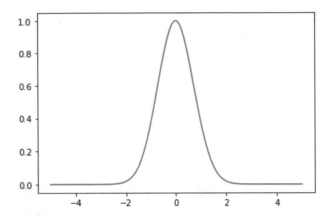

Figure 2-7. *Visualizing the squared exponential kernel function*

Note that $K_{ii} = \kappa(x_i, x_i) = 1$ at every point $x_i \in A$. Thus, the covariance between two points x_i and x_i also measures the correlation between the corresponding functional values $f(x_i)$ and $f(x_j)$ based on the definition of correlation introduced earlier. Assuming a mean function that returns zero across the whole domain, we can see that the marginal distribution of the random variable at every input location follows a standard normal distribution.

From another perspective, the kernel function can also be considered to determine the height of the three random variables $\{f_1, f_2, f_3\}$, which represents our entity of interest, given a fixed set of locations $\{x_1, x_2, x_3\}$. We can also introduce additional parameters to the kernel function so that it becomes more flexible in characterizing the particular dataset, which could be in the form of images or texts.

Extending to Other Variables

Suppose we would like to obtain the distribution of another variable in terms of its mean and variance at other input locations given the observed dataset D. Quantifying the distribution of other variables based on the observed ones is a key characteristic of GP. To this end, we can construct a joint distribution of the observed set of variables together with the new variable and apply the same technique to get the conditional distribution for the new variable. Let us build on the previous example.

Given a new input location x_*, we can pick three possible values/realizations (high, medium, and low) for the corresponding random variable f_* (conditionally independent from all other variables in $\mathbf{f}$), each assuming a different probability of occurrence. As illustrated in Figure 2-8, given that closer points will be similar in value as encoded by the kernel function, we can intuit that the middle point has the highest probability, which has roughly the same distance with the second and third observations. That is, a small change in the input location should result in a small change in the output observation value. Choosing the middle would result in a smooth curve when connecting it with other observations. In addition, if we were to move x_* closer to x_2, we would expect f_* to increase in value and further approach f_2.

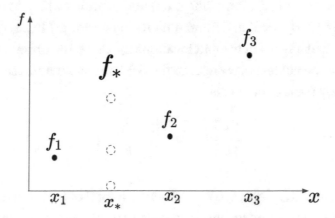

Figure 2-8. *Three possible values of f_* (dash circles) for a new input location x_*. Based on the smoothness assumption encoded in the kernel/covariance function, we would intuit the middle point to be the most likely candidate*

Let us analyze how this intuition can be incorporated using the GP framework. We can model the corresponding observation as a random variable f_* that follows a joint Gaussian distribution together with the observed random variables $\mathbf{f} = \{f_1, f_2, f_3\}$. Lumping all variables together allows us to enforce the similarity constraints as encoded in the covariance function.

Specifically, the observed variables $\mathbf{f}$ and the unobserved variable f_* are respectively distributed as follows:

$$\mathbf{f} \sim N(\mathbf{0}, \mathbf{K})$$

$$f_* \sim N(0, \kappa(x_*, x_*))$$

where the self-covariance term $K(x_*, x_*)$ is just the variance of x_*, which would equal 1 when using the squared exponential kernel function. To make the setting more general, we assume a finite number of random variables in $\mathbf{f}$ observed at locations $\mathbf{x}$ without noise and jointly following a multivariate Gaussian distribution:

$$p(\mathbf{f}|\mathbf{x}) = N(\mathbf{f}|,\mathbf{0}|,\mathbf{K})$$

which is the same as $\mathbf{f} \sim N(\mathbf{0}, \mathbf{K})$. Revealing these observations will help update our prior belief $p(\mathbf{f}_*|\mathbf{x}_*) = N(\mathbf{f}_*|\mathbf{0}, \kappa(\mathbf{x}_*, \mathbf{x}_*))$ of GP and obtain a posterior $p(\mathbf{f}_*|\mathbf{x}_*, \mathbf{x}, \mathbf{f})$ for a set of new random variables $\mathbf{f}_*$ at a set of new input locations $\mathbf{x}_*$, also to be observed without noise. To see how this can be achieved, we refer to the GP framework and use the fact that $\mathbf{f}$ and $\mathbf{f}_*$ will also be jointly Gaussian. That is

$$\begin{bmatrix} \mathbf{f} \\ \mathbf{f}_* \end{bmatrix} \sim N\left(\mathbf{0}, \begin{bmatrix} \mathbf{K} & \mathbf{K}_* \\ \mathbf{K}_*^T & \mathbf{K}_{**} \end{bmatrix} \right)$$

where $\mathbf{K}_* = \kappa(\mathbf{x}, \mathbf{x}_*)$ and $\mathbf{K}_{**} = \kappa(\mathbf{x}_*, \mathbf{x}_*)$. When there is a total of n observed samples (i.e., training data) and n_* new input locations, we would have a $n \times n$ matrix for $\mathbf{K}$, a $n \times n_*$ matrix for $\mathbf{K}_*$, and a $n_* \times n_*$ matrix for $\mathbf{K}_{**}$.

With $\mathbf{f}$ and $\mathbf{f}_*$ modeled as a joint Gaussian distribution, we can again rely on the multivariate Gaussian theorem and directly use the closed-form solution to obtain the parameters of the conditional distribution $p(\mathbf{f}_*|\mathbf{x}_*, \mathbf{x}, \mathbf{f})$:

$$p(\mathbf{f}_*|\mathbf{x}_*, \mathbf{x}, \mathbf{f}) = N(\mathbf{f}_*|\mu_*, \Sigma_*)$$

$$\mu_* = \mathbf{K}_*^T \mathbf{K}^{-1} \mathbf{f}$$

$$\Sigma_* = \mathbf{K}_{**} - \mathbf{K}_*^T \mathbf{K}^{-1} \mathbf{K}_*$$

The updated posterior $p(\mathbf{f}_*|\mathbf{x}_*, \mathbf{x}, \mathbf{f})$ with such parameters will thus assign the highest probability to the middle point (mean estimate) among the three candidates in Figure 2-8.

Figure 2-9 summarizes the derivation process of learning from existing data and obtaining the predictive (conditional) distribution for the new data under the GP framework. The key part is defining the kernel function as a form of a prior belief or an inductive bias and iteratively refining it by incorporating information from the observations.

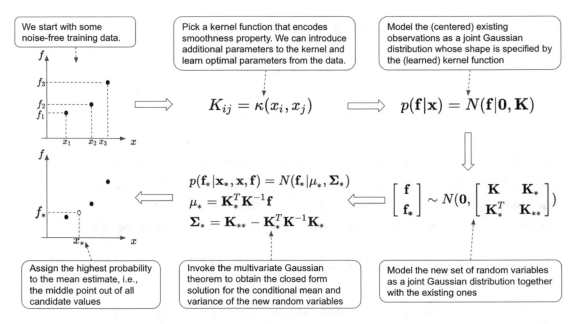

Figure 2-9. *Obtaining the predictive distribution of new input locations using the Gaussian process. We start by defining and optionally optimizing a kernel function to encode what is observed from actual samples and what is believed using the prior. Then we model all variables as jointly Gaussian and obtain the conditional distribution for the new variables*

Learning from Noisy Observations

In our running example, we assumed that the sample observations were noise-free. In practice, such observations are often noise-corrupted, adding another layer of uncertainty to the modeling. A common treatment is to model the noises for each observation as conditionally independent random variables and additive to the noise-free observations. In other words, we can represent the actual observations using $\mathbf{y} = \mathbf{f} + \boldsymbol{\varepsilon}$, where $\boldsymbol{\varepsilon} \sim N\left(\mathbf{0}, \sigma_y^2 \mathbf{I}\right)$ with a homogenous variance. However, we can also model the case with heterogeneous variance for each input location.

The biggest benefit of using an additive Gaussian noise is that we could perform the same inference exactly, obtaining a closed-form solution for the parameters of the updated posterior Gaussian process. Following the same derivation process, the predictive distribution for the noise-free predictions $\mathbf{f}_*$ now becomes

$$p\left(\mathbf{f}_*|\mathbf{x}_*,\mathbf{x},\mathbf{y}\right)=N\left(\mathbf{f}_*|\mu_*,\Sigma_*\right)$$

$$\mu_* = \mathbf{K}_*^T\mathbf{K}_y^{-1}\mathbf{y}$$

$$\Sigma_* = \mathbf{K}_{**} - \mathbf{K}_*^T\mathbf{K}_y^{-1}\mathbf{K}_*$$

where $\mathbf{K}_y = \mathbf{K}+\sigma_y^2\mathbf{I}$ is the variance of the actual noise-corrupted samples, which absorbs the respective variance from the true noise-free observation and the noise. Since the noise will also stick around for each random variable in $\mathbf{f}_*$, we can additionally incorporate the noises $\boldsymbol{\varepsilon}$ in the noise-corrupted predictions $\mathbf{y}_*$ by adding $\sigma_y^2\mathbf{I}$ to Σ_*:

$$p\left(\mathbf{y}_*|\mathbf{x}_*,\mathbf{x},\mathbf{y}\right)=N\left(\mathbf{y}_*|\mu_*,\Sigma_*+\sigma_y^2\mathbf{I}\right)$$

We can then use this predictive distribution to perform sampling at any input location within the domain. When the number of samples grows to infinity, connecting these samples would form a functional curve, which is a sample from the Gaussian process. We can also plot the confidence interval by connecting the pointwise standard deviations. Viewing the functional curve as a composition of infinitely many pointwise samples also follows the same argument as the case of sampling from a bivariate Gaussian distribution.

Gaussian Process in Practice

In this section, we will go through a concrete example of drawing functional curves as samples from a Gaussian process using its prior distribution, updating the parameters and obtaining refreshed samples using its posterior distribution. We will use Python to illustrate the implementation steps.

Drawing from GP Prior

Let us look at how to sample functional curves from the prior of a Gaussian process before revealing any observations. We will follow the following steps, which is similar to sampling from an arbitrary Gaussian distribution:

- Create a list of n candidate input locations $\{x_i, i = 1, ..., n\}$. We are using discrete locations X_test to approximate a continuous domain. When $n \to \infty$, we would obtain a functional curve over the infinitely many input locations.

- Initialize the mean vector $\boldsymbol{\mu}$ (with n elements) and the covariance matrix $\mathbf{K}$ (with n x n elements). We assume the data is centered and thus $\boldsymbol{\mu} = \mathbf{0}$, and $\mathbf{K}$ is a squared exponential kernel function calculated and stored in variable K.

- Perform the Cholesky decomposition $\mathbf{K} = \mathbf{LL}^T$ to obtain $\mathbf{L}$, which is stored in variable L.

- Obtain a sample over $N(\mathbf{0}, \mathbf{K})$ via $\mathbf{L}N(\mathbf{0}, \mathbf{I})$ and stored in f_prior.

See Listing 2-2 for the full code on generating samples from a GP prior.

Listing 2-2. Drawing samples from the GP prior

```
# import necessary packages
import numpy as np
from matplotlib import pyplot as plt
%matplotlib inline
# set random seed to ensure reproducibility
np.random.seed(8)

# define a kernel function to return a squared exponential distance between
two input locations
def kernel(a, b):
    # decomposing the squaring operation into three parts
    # each input location may be multi-dimensional, thus summing over all
      dimensions
    sq_dist = np.sum(a**2,1).reshape(-1,1) + np.sum(b**2,1) - 2*np.
    dot(a,b.T)
    return np.exp(-sq_dist)

# setting number of input locations which approximates a function when
growing to infinity
n = 100
X_test = np.linspace(-5,5,n).reshape(-1,1)
```

```
# calculate the pairwise distance, resulting in a nxn matrix
K = kernel(X_test, X_test)

# adding a small number along diagonal elements to ensure cholesky
decomposition works
L = np.linalg.cholesky(K + 1e-10*np.eye(n))
# calculating functional samples by multiplying the sd with standard
normal samples
samples = np.dot(L,np.random.normal(size=(n,5)))
plt.plot(X_test, samples)
```

Running the preceding code will generate five random curves, each consisting of 100 discrete points connected to approximate a function across the whole domain. See Figure 2-10 for the output.

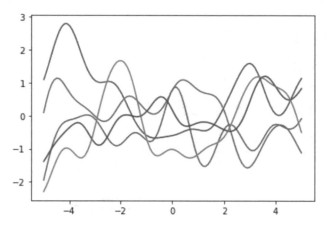

Figure 2-10. *Drawing five random samples from a GP prior*

We can make the kernel function more flexible by introducing additional parameters. For example, the widely used *Gaussian kernel* or *RBF kernel* uses two parameters: the length parameter l to control the smoothness of the function and σ_f to control the vertical variation:

$$\kappa\left(x_i,x_j\right)=\sigma_f^2\exp\left(-\frac{1}{2l^2}\|x_i-x_j\|^2\right)$$

For demonstration purposes, we will use the same parameters for each input location/dimension, which also results in the so-called *isotropic kernel* and is defined as follows. Both input arguments could include multiple elements/locations, each consisting of one or more dimensions. For example, if **X1** is a (m x d) matrix with m locations and d dimensions for each location, and **X2** is (n x d) matrix, applying the kernel function would result in a (m x n) matrix containing pairwise similarity entries.

Listing 2-3. Defining the isotropic squared exponential kernel

```
# Args:
#      X1: array of m points (m x d).
#      X2: array of n points (n x d).
# Returns:
#      (m x n) matrix.
def ise_kernel(X1, X2, l=1.0, sigma_f=1.0):
    sq_dist = np.sum(X1**2,1).reshape(-1,1) + np.sum(X2**2,1) - 2*np.
    dot(X1,X2.T)
    return sigma_f**2 * np.exp(-0.5/l**2 * sq_dist)
```

We can obtain the mean vector (assumed to be zero) and covariance matrix (using $l = 1$ and $\sigma_f = 1$) based on a list of input locations previously defined in X_test.

Listing 2-4. Calculating the mean vector and covariance matrix

```
# mean and covariance of the prior
mu = np.zeros(X_test.shape)
K = ise_kernel(X_test, X_test)
```

Previously, we drew from the multivariate Gaussian distribution by scaling samples from a standard normal distribution. In fact, there is a function multivariate_normal from numpy to help us achieve this. We draw five samples as follows.

Listing 2-5. Drawing from multivariate Gaussian distribution

```
# draw samples from the prior using multivariate_normal from numpy
# convert mu from shape (n,1) to (n,)
samples = np.random.multivariate_normal(mean=mu.ravel(), cov=K, size=5)
```

We can then plot the five samples together with the mean function and 95% confidence bounds that form the uncertainty region based on the diagonal entries of the covariance matrix (contained in the uncertainty variable). We define a utility function to plot the samples across the predefined grid values in the following code listing. The function also has two input arguments (X_train and Y_train) as placeholders when actual observations are revealed.

Listing 2-6. Plotting GP prior mean function, uncertainty region, and samples

```
def plot_gp(mu, cov, X, X_train=None, Y_train=None, samples=[]):
    X = X.ravel()
    mu = mu.ravel()
    # 95% confidence interval
    uncertainty = 1.96 * np.sqrt(np.diag(cov))

    plt.fill_between(X, mu + uncertainty, mu - uncertainty, alpha=0.1)
    plt.plot(X, mu, label='Mean')
    for i, sample in enumerate(samples):
        plt.plot(X, sample, lw=1, ls='--', label=f'Sample {i+1}')
    # plot observations if available
    if X_train is not None:
        plt.plot(X_train, Y_train, 'rx')
    plt.legend()
```

We can then supply the necessary ingredients to invoke this function, which produces an output shown in Figure 2-11:

```
>>> plot_gp(mu, K, X_test, samples=samples)
```

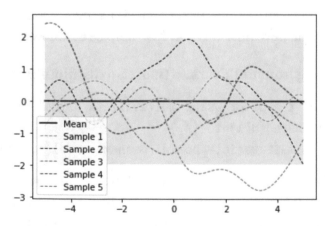

Figure 2-11. *Five samples from the GP prior together with the mean function and 95% confidence intervals*

Obtaining GP Posterior with Noise-Free Observations

After observing some samples in the form of input location and output functional value pairs, we can treat them as training data to refresh the GP prior and obtain an updated posterior specified by its mean function and covariance function. Since we are using discrete locations contained in a dense grid to approximate a continuous domain, the posterior parameters of the GP will be represented via a mean vector and a covariance matrix.

In the following code listing, we define a function to compute the parameters of the posterior GP using the training data X_train and Y_train, the new input locations X_s, kernel-related parameters l and sigma_f, and standard deviation of the additive Gaussian noise sigma_y. The respective entries in the covariance matrix of the joint multivariate Gaussian distribution between existing inputs X_train and new inputs X_s are first computed and stored in K, K_s, and K_ss, respectively, followed by plugging in the closed-form formula to obtain the posterior mean vector mu_s and covariance matrix cov_s for the new inputs.

Listing 2-7. Calculating GP posterior mean vector and covariance matrix for the new inputs

```
from numpy.linalg import inv

def update_posterior(X_s, X_train, Y_train, l=1.0, sigma_f=1.0, sigma_
y=1e-8):
    """
    Computes the mean vector and covariance matrix of the posterior
    distribution
    from m training data X_train and Y_train and n new inputs X_s.
    Args:
        X_s: New input locations (n x d).
        X_train: Training locations (m x d).
        Y_train: Training targets (m x 1).
        l: Kernel length parameter.
        sigma_f: Kernel vertical variation parameter.
        sigma_y: Observation noise parameter.
```

```
    Returns:
        Posterior mean vector (n x d) and covariance matrix (n x n).
    """
    # covariance matrix for observed inputs
    K = ise_kernel(X_train, X_train, l, sigma_f) + sigma_y**2 *
    np.eye(len(X_train))
    # cross-variance between observed and new inputs
    K_s = ise_kernel(X_train, X_s, l, sigma_f)
    # covariance matrix for new inputs
    K_ss = ise_kernel(X_s, X_s, l, sigma_f) + 1e-8 * np.eye(len(X_s))
    # computer inverse of covariance matrix
    K_inv = inv(K)
    # posterior mean vector based on derived closed-form formula
    mu_s = K_s.T.dot(K_inv).dot(Y_train)
    # posterior covariance matrix based on derived closed-form formula
    cov_s = K_ss - K_s.T.dot(K_inv).dot(K_s)

    return mu_s, cov_s
```

Now we can pass in a few noise-free training data points and apply the function to obtain posterior mean and covariance and generate updated samples from the GP posterior.

Listing 2-8. Generating samples from posterior GP

```
# noise free training data
X_train = np.array([-4, -3, -2, -1, 1]).reshape(-1, 1)
Y_train = np.cos(X_train)
# compute mean and covariance of the posterior distribution
mu_s, cov_s = update_posterior(X_test, X_train, Y_train)
# generate five samples from multivariate normal distribution
samples = np.random.multivariate_normal(mu_s.ravel(), cov_s, 5)
```

We can call the same plotting function again with the training data passed in as additional parameters and plotted as crosses.

```
>>> plot_gp(mu_s, cov_s, X_test, X_train=X_train, Y_train=Y_train,
samples=samples)
```

Running the preceding code produces Figure 2-12. The posterior mean function smoothly interpolates the observed inputs where the corresponding variances of the random variables are also zero. As represented by the confidence interval, the uncertainty is small in the neighborhood of observed locations and increases as we move further away from the observed inputs. In the meantime, the marginal distributions at locations sufficiently far away from the observed locations will largely remain unchanged, given that the prior covariance function essentially encodes no correlation for distant locations.

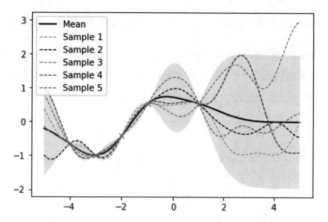

Figure 2-12. *Drawing five samples from GP posterior based on noise-free observations*

Working with Noisy Observations

When the observations are noisy, the total variance of an observed input will stem from both the uncertainty of the random variable at the input location as well as the variance of the noise. As mentioned earlier, the noise often arises due to measurement error or inexact statistical approximation. We can estimate the noise-free covariance matrix $\mathbf{K}$ at the observed inputs as usual and add the variance of the noise to the diagonal entries (self-covariance) to obtain the noise-corrupted covariance matrix $\mathbf{K}_y = \mathbf{K} + \sigma_y^2 \mathbf{I}$. See the following code listing for an illustration.

Listing 2-9. Obtaining GP posterior with noisy observations

```
# set standard deviation of the noise
noise = 0.5
# create noisy training data
Y_train = np.cos(X_train) + noise * np.random.randn(*X_train.shape)
# compute mean and covariance of the posterior distribution
mu_s, cov_s = update_posterior(X_test, X_train, Y_train, sigma_y=noise)
# generate five samples
samples = np.random.multivariate_normal(mu_s.ravel(), cov_s, 5)
```

Running the preceding code generates Figure 2-13, where the mean function does not interpolate the actual training observations due to the extra noise term in $\mathbf{K}_y$ and the variances at the corresponding input locations are also nonzero.

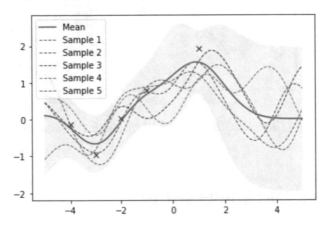

Figure 2-13. *Drawing five samples from GP posterior based on noisy observations*

In the previous example, we set the standard deviation of the noise term (assumed to be independent of the underlying random variable of interest) as 0.5. Increasing the noise variance would lead to further deviation of actual observation from the underlying true functional value at the specific input location, resulting in additional inflation of the confidence interval across the whole domain. As verified in Figure 2-14, by setting noise=1, the confidence interval becomes wider given a lower signal-to-noise ratio.

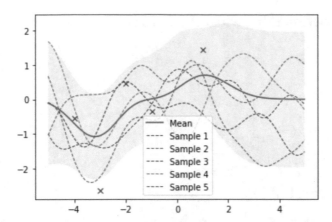

Figure 2-14. *A wider confidence interval across the whole input domain given a lower signal-to-noise ratio*

Also, note that the fitted mean function becomes smoother when given a stronger noise. By making coarser approximations, noisier training data thus helps prevent the model from becoming too wiggly and, as a result, prone to overfitting.

Experimenting with Different Kernel Parameters

Earlier, we have adopted an isotropic kernel with $l = 1$ and $\sigma_f = 1$ for all observation locations. l controls the smoothness of the GP fit. A higher l will lead to a smoother fit with less curvature, while a lower l will end up with a more flexible fit with wider uncertainty intervals. σ_f controls the vertical variation of the functional samples drawn from GP. A higher σ_f will result in wider uncertainty regions across the domain as we move away from the observed location.

In the following code listing, four combinations of l and σ_f are selected to demonstrate the impact of different parameter settings on the resulting fit. These combinations are stored as a list of tuples in params and iterated in a loop to generate four figures arranged in two rows.

Listing 2-10. Experimenting with different kernel parameters

```
params = [
    (0.1, 1.0),
    (2.0, 1.0),
    (1.0, 0.1),
    (1.0, 2.0)
]

plt.figure(figsize=(12, 5))

for i, (l, sigma_f) in enumerate(params):
    mu_s, cov_s = update_posterior(X_test, X_train, Y_train, l=l,
                            sigma_f=sigma_f)
    plt.subplot(2, 2, i + 1)
    plt.subplots_adjust(top=2)
    plt.title(f'l = {l}, sigma_f = {sigma_f}')
    plot_gp(mu_s, cov_s, X_test, X_train=X_train, Y_train=Y_train)
```

The output of running the preceding code is shown in Figure 2-15. In the top row, increasing l while controlling σ_f from the left to the right plot shows that the resulting fit is less wiggly and associated with more narrow uncertainty regions. When fixing l and increasing σ_f from the left to the right plot in the bottom row, we observe a much wider uncertainty region, especially toward the locations at the right end.

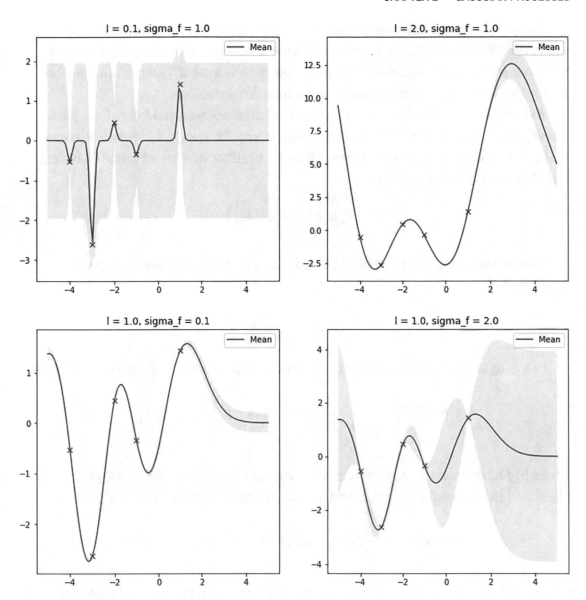

Figure 2-15. *Illustrating four different settings of kernel parameters*

Hyperparameter Tuning

The kernel parameters l and σ_f are fixed before fitting a GP and are often referred to as the hyperparameters. In the last section, we showed the impact of different *manually* set hyperparameters on the resulting fit. Instead of testing over multiple different combinations one by one, a better approach is to automatically identify the value of the hyperparameters based on the characteristics of the observed data under the GP framework. Since we already have access to the marginal likelihood of these observations, a common approach is to go with the set of hyperparameters that maximize the joint log-likelihood of these marginal distributions.

Assume a total of n noisy samples $\mathbf{y}$ observed at locations $\mathbf{x}$ and following a joint Gaussian distribution with mean vector $\boldsymbol{\mu}$ and covariance matrix $\mathbf{K}_y$. We will also use $\theta = \{l, \sigma_f\}$ to denote all hyperparameters. The joint likelihood of the marginal Gaussian distribution can be expressed as

$$p(\mathbf{y}|\mathbf{x},\theta) = N\left(\mathbf{y}|\boldsymbol{\mu},\mathbf{K}_y\right)$$

We can then obtain the optimal hyperparameters $\hat{\theta}$ by maximizing the joint likelihood:

$$\hat{\theta} = \operatorname{argmax}_\theta p(\mathbf{y}|\mathbf{x},\theta)$$

Plugging in the definition of a multivariate Gaussian distribution gives the following:

$$p(\mathbf{y}|\mathbf{x},\theta) = \frac{1}{\sqrt{(2\pi)^n |\mathbf{K}_y|}} \exp\left(-\frac{1}{2}(\mathbf{y}-\boldsymbol{\mu})^T \mathbf{K}_y^{-1}(\mathbf{y}-\boldsymbol{\mu})\right)$$

where $|\mathbf{K}_y|$ represents the determinant of $\mathbf{K}_y$. We can then take the log of the joint likelihood to remove the exponent and instead maximize the log likelihood:

$$\log p(\mathbf{y}|\mathbf{x},\theta) = -\frac{1}{2}(\mathbf{y}-\boldsymbol{\mu})^T \mathbf{K}_y^{-1}(\mathbf{y}-\boldsymbol{\mu}) - \frac{n}{2}\log 2\pi - \frac{1}{2}\log |\mathbf{K}_y|$$

where the first term measures the goodness of fit, and the last two terms serve as a form of entropy-based penalty. We can further simplify the formula by assuming $\boldsymbol{\mu} = \mathbf{0}$, resulting in

$$\log p(\mathbf{y}|\mathbf{x},\theta) = -\frac{1}{2}\mathbf{y}^T\mathbf{K}_y^{-1}\mathbf{y} - \frac{n}{2}\log 2\pi - \frac{1}{2}\log |\mathbf{K}_y|$$

In practice, the optimal hyperparameters $\hat{\theta}$ are found by minimizing the *negative log-likelihood* (NLL) of the marginal distributions:

$$\hat{\theta} = \text{argmin}_\theta - \log p\left(\mathbf{y}|\mathbf{x},\theta\right)$$

Let us look at how to implement this. In the following code listing, we define a function that takes in the training data and noise level and returns a function that calculates the negative log marginal likelihood based on the specified parameters (l and σ_f). This is a direct implementation based on the definition of the negative log-likelihood earlier.

Listing 2-11. Calculating negative log marginal likelihood

```
from numpy.linalg import det
from scipy.optimize import minimize

# direct implementation, may be numerically unstable
def nll_direct(X_train, Y_train, noise) :
    """
    Returns a function that computes the negative log marginal
    likelihood for training data X_train and Y_train and given
    noise level.

    Args:
        X_train: training locations (m x d).
        Y_train: training targets (m x 1).
        Noise: known noise level of Y_train.

    Returns:
        Minimization objective function.
    """
    Y_train = Y_train.ravel()

    def nll(theta):
        K = ise_kernel(X_train, X_train, l=theta[0], sigma_f=theta[1]) + \
            noise**2 * np.eye(len(X_train))
        return 0.5 * Y_train.dot(inv(K).dot(Y_train)) + \
               0.5 * len(X_train) * np.log(2*np.pi) + \
               0.5 * np.log(det(K))

    return nll
```

We can invoke the `minimize` function from `numpy` to perform the minimization using L-BFGS-B, a widely used optimization algorithm. The optimization procedure starts with both l and σ_f equal to one and searches for the optimal parameters. In practice, multiple rounds of searches with different starting points are often conducted to avoid the same local minima.

Listing 2-12. GP posterior using optimal kernel parameters

```
# initialize at [1,1] and search for optimal values within between 1e-5 to
infinity
res = minimize(nll_direct(X_train, Y_train, noise), [1, 1],
               bounds=((1e-5, None), (1e-5, None)),
               method='L-BFGS-B')
l_opt, sigma_f_opt = res.x
# compute posterior mean and covariance with optimized kernel parameters
and plot the results
mu_s, cov_s = update_posterior(X_test, X_train, Y_train, l=l_opt,
                               sigma_f=sigma_f_opt, sigma_y=noise)
plot_gp(mu_s, cov_s, X_test, X_train=X_train, Y_train=Y_train)
```

Running the preceding code will generate Figure 2-16, where the training data points are reasonably covered with the 95% uncertainty regions. The mean function of the GP posterior also appears to be a good approximation.

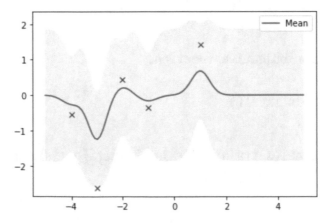

Figure 2-16. *GP posterior using optimized kernel parameters*

Note that directly calculating the negative log marginal likelihood based on the given formula may be numerically unstable and computationally expensive as the problem starts to scale. In particular, calculating $\mathbf{K}_y^{-1}$ involves inverting the matrix, which becomes computationally costly if the matrix grows. In practice, a neat trick is to convert the matrix inversion to another more computation-friendly operation and indirectly obtain the same result.

To put things in perspective, let us look at the first term $-\dfrac{1}{2}\mathbf{y}^T\mathbf{K}_y^{-1}\mathbf{y}$ in $\log p(\mathbf{y}|\mathbf{x},\theta)$.

Calculating $\mathbf{K}_y^{-1}$ using inv(K) may cause numerical instability as in the previous code. To circumvent the direct calculation, we can treat the calculation of $\mathbf{K}_y^{-1}\mathbf{y}$ as a whole and use the Cholesky decomposition trick to solve a system of equations to avoid direct matrix inversion. Specifically, we note that $\mathbf{K}_y^{-1}\mathbf{y} = \mathbf{L}^{-T}\mathbf{L}^{-1}\mathbf{y}$, where we have used the Cholesky decomposition to convert $\mathbf{K}_y$ into the product of a triangular factor matrix $\mathbf{L}$ and its transpose $\mathbf{L}^T$, that is, $\mathbf{K}_y = \mathbf{LL}^T$. The triangular matrix is now much more memory efficient when performing mathematical operations, and only one triangular matrix needs to be computed and stored.

Assuming $\mathbf{Lm} = \mathbf{y}$, then we can easily solve for $\mathbf{m} = \mathbf{L}^{-1}\mathbf{y}$ using the solve_triangular function from the scipy package. Similarly, assuming $\mathbf{L}^T\alpha=$, we can use the same function to solve for $\alpha = \mathbf{L}^{-T}\mathbf{m} = \mathbf{L}^{-T}\mathbf{L}^{-1}\mathbf{y}$. Therefore, we can calculate $\mathbf{K}_y^{-1}\mathbf{y}$ by solving two systems of equations based on the result from the Cholesky decomposition.

In addition, we can further accelerate the calculation by using the fact that

$\log|\mathbf{K}_y| = 2\sum_{i=1}^{n}\log L_{ii}$, where L_{ii} is the ith diagonal entry of factor matrix $\mathbf{L}$. The preceding

reformulations are implemented in the following code listing.

Listing 2-13. More numerically stable and faster implementation in calculating NLL

```
from numpy.linalg import cholesky
from scipy.linalg import solve_triangular

# indirect implementation which is more numerically stable
def nll_stable(X_train, Y_train, noise):
    Y_train = Y_train.ravel()
```

```
def nll(theta):
    K = ise_kernel(X_train, X_train, l=theta[0], sigma_f=theta[1]) + \
        noise**2 * np.eye(len(X_train))
    L = cholesky(K)
    # solve the first system of equations
    m = solve_triangular(L, Y_train, lower=True)
    # solve the second system of equations
    alpha = solve_triangular(L.T, m, lower=False)

    return 0.5 * Y_train.dot(alpha) + \
           0.5 * len(X_train) * np.log(2*np.pi) + \
           np.sum(np.log(np.diagonal(L)))

return nll
```

Running the preceding code will generate the same plot as earlier, but the difference reveals when the problem becomes large. To see the full list of implementations, refer to the accompanying notebook at `https://github.com/jackliu333/bayesian_optimization_theory_and_practice_with_python/blob/main/Chapter_2.ipynb`.

Summary

A Gaussian process is an important tool in many modeling problems. In this chapter, we covered the following list of items:

- A Gaussian process is characterized by a mean function and a covariance function. It extends the case of finite multivariate Gaussian distribution to an infinite case.

- The kernel/covariance function governs the shape of the fitted curves such as smoothness, introduced as a form of inductive bias.

- Sampling from an arbitrary (independent) multivariate Gaussian distribution can be performed by sampling from a normal distribution followed by the scale-location transformation.

- Gaussian process regression treats the observed random variables as joint multivariate Gaussian and provides the closed-form predictive marginal distribution for the new input locations using the multivariate Gaussian theorem.

- Noisy observations with additive and normally distributed noise can be seamlessly incorporated in the GP framework.

- Implementing a GP learning process starts with calculating the parameters for the GP prior and updating them to obtain the GP posterior using observed samples, where the mean function of the GP posterior would interpolate the noise-free observations.

- The kernel parameters can be optimized using the maximal likelihood estimation procedure, where the implementation can be made more numerically stable and faster using the Cholesky decomposition.

In the next chapter, we will look at expected improvement (EI), the most widely used acquisition function in Bayesian optimization.

CHAPTER 3

Bayesian Decision Theory and Expected Improvement

The previous chapter used Gaussian processes (GP) as the surrogate model to approximate the underlying objective function. GP is a flexible framework that provides uncertainty estimates in the form of probability distributions over plausible functions across the entire domain. We could then resort to the closed-form posterior predictive distributions at proposed locations to obtain an educated guess on the potential observations.

However, it is not the only choice of surrogate model used in Bayesian optimization. Many other models, such as random forest, have seen increasing use in recent years, although the default and mainstream choice is still a GP. Nevertheless, the canonical Bayesian optimization framework allows any surrogate model as long as it provides a posterior estimate for the function, which then gets used by the acquisition function to generate a sampling proposal.

The acquisition function bears even more choices and is an increasingly crowded research space. Standard acquisition functions such as expected improvement and upper confidence bound have seen wide usage in many applications, and problem-specific acquisition functions incorporating domain knowledge, such as safe constraint, are constantly being proposed. The acquisition function assumes a more important role in the Bayesian optimization framework as it directly determines the sampling decision for follow-up data acquisition. A good acquisition function thus enables the optimizer to locate the (global) optimum as fast as possible, where the optimum is measured in the sense of the location that holds the optimum value or the optimum value across the whole domain.

69

© Peng Liu 2023
P. Liu, *Bayesian Optimization*, https://doi.org/10.1007/978-1-4842-9063-7_3

This chapter will dive into the Bayesian optimization pipeline using expected improvement, the most widely used acquisition function for sampling decisions. We will first characterize Bayesian optimization as a sequential decision process under uncertainty, followed by a thorough introduction of expected improvement. An intelligent selection of the following sampling location that involves uncertainty estimates is the key to achieving sample-efficient global optimization. Lastly, we will go through a case study using expected improvement to guide hyperparameter tuning.

Optimization via the Sequential Decision-Making

The Bayesian optimization framework sequentially determines the following sampling location in search of the global optimum, usually assumed to be maximum in a maximization setting. Based on a specific policy, the optimizer would collect observed data points, update the posterior belief about the probability distributions of the underlying functions, propose the next sampling point for probing, and finally collect the additional data point at the proposed location and repeat. This completes one iteration within the iterative and sequential process. Our knowledge of the underlying function constantly evolves and gets updated every time a new data point is incorporated under the guidance of the existing collection of observations.

At the end of the process, the policy will return either the location of the optimal value or the optimum itself, although we are often interested in the optimal location. This is often referred to as the *outer loop* in the Bayesian optimization framework. Besides, maximizing the acquisition function to generate the following sampling location constitutes the *inner loop* of Bayesian optimization. The acquisition function serves as a side computation within the inner loop to aid the subsequent sampling decision. Optimizing the acquisition function is usually considered fast and cheap due to its inexpensive evaluation and analytical differentiability. We can obtain the closed-form expression for some acquisition functions and access its gradient by using an off-the-shelf optimization procedure. We can also resort to approximation methods such as Monte Carlo estimation for more complex acquisition functions without closed-form expression.

In addition, the policy would also need to consider *when* to terminate the probing process, especially in the case of a limited probing budget. Upon termination, the optimizer would return the optimal functional value or location, which may or may not live in the observed locations and could exist anywhere within the domain.

The optimizer thus needs to trade off between calling off the query and performing additional sampling, which incurs an additional cost. Therefore, the action space of the optimizer contains not only the sampling location but also a binary decision on termination.

Figure 3-1 characterizes the sequential decision-making process that underpins Bayesian optimization. The policy would propose the following sampling location at each outer loop iteration or terminate the loop. Suppose it decides to propose an additional sampling action. In that case, we will enter the inner loop to seek the most promising location with the highest value of the prespecified acquisition function. We would then probe the most favorable location and append the additional observation in our data collection, which is then used to update the posterior belief on the underlying objective function through GP. On the other hand, if the policy believes the additional query is not worth the corresponding cost to improve our belief on the global optimum, it would decide to terminate the outer loop and return the current best estimate of the global optimum or its location. This also forms the *stopping rule* of the policy, which could be triggered upon exhausting the limited budget or assuming an adaptive mechanism based on the current progression.

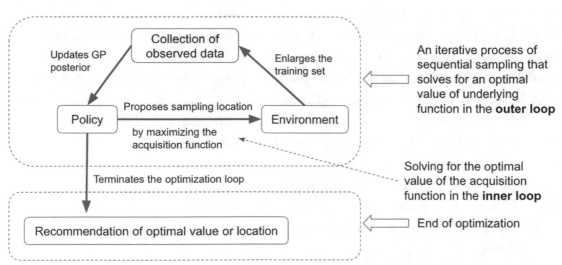

Figure 3-1. *Sequential decision-making process in Bayesian optimization. The outer loop performs optimization in search of the global optimum by sequential sampling across the entire domain. Each iteration is based on the output of the inner loop, which involves another separate optimization*

Quantifying the improvement on the belief of the global optimum is reflected in the *expected marginal gain* on the *utility* of observed data, which is the core concept in the Bayesian decision theory used in Bayesian optimization. We will cover this topic in the following sections.

Seeking the Optimal Policy

The ultimate goal of BO is to develop an intelligent policy that performs sequential decision-making under uncertainty in a principled manner. When the policy is measured in terms of the quality of the collected data, BO would then seek the *optimal policy* that maximizes the expected quality of the collected data. In other words, the optimal policy would deliver the most informative dataset on average to assist the task of locating the global optimum while considering the posterior belief about the underlying objective function.

In this regard, the acquisition function is used to measure the data quality when considering the following sampling decision across the entire domain. The acquisition function maps each candidate location to a numeric score, which essentially encodes preferences over different candidate locations. It serves as the intermediate calculation that bridges the gap between updating the posterior belief and seeking the optimal policy. Specifically, the optimal decision based on the most updated posterior belief is made by choosing the location with the maximal score calculated using the specified acquisition function, which completes one round of inner optimization.

Mathematically, for each candidate location x in domain A, the self-defined acquisition function $\alpha(x)$ maps each x to a scalar value, that is, $\alpha : A \to \mathbb{R}$. Here, we assume x is single-dimensional without loss of generality and for the seek of notational convenience, although it can assume multiple features, that is, $\mathbf{x} \in \mathbb{R}^d$. Besides, the acquisition function is an evolving scoring function that also depends on the currently collected dataset $\mathcal{D}_n$ with n observations, thus writing $\alpha(x; \mathcal{D}_n)$ to indicate such dependence.

Therefore, for any arbitrary locations x_1 and x_2 within A, we would prefer x_1 over x_2 if $\alpha(x_1; \mathcal{D}_n) > \alpha(x_2; \mathcal{D}_n)$, and vice versa. Out of infinitely many candidate locations in the case of a continuous search domain, the optimal policy would then act greedily by selecting the single location x_{n+1}^* with the highest acquisition value as the following sampling action. We use the superscript $*$ to denote an optimal action and the subscript $n + 1$ to indicate the additional first future sampling location on top of the existing n observations. The addition of one thus means the lookahead horizon or time step into the future. The notation x_{n+1} denotes all possible candidate locations, including the observed ones, and is viewed as a random variable. The optimal decision is then defined as follows:

$$x_{n+1}^* = \text{argmax}_{x_{n+1} \in A}\, \alpha\left(x_{n+1}; \mathcal{D}_n\right)$$

Common acquisition functions such as expected improvement (to be introduced later) admit fast gradient-based optimization due to the availability of the closed-form analytic expression and the corresponding gradient. This means that we have converted the original quest for global optimization of a difficult and unknown objective function to a series of fast optimizations of a known acquisition function. However, as we will learn later, some acquisition functions, especially those featuring multi-step lookahead, may not be analytically differentiable, making the inner optimization a nontrivial problem. In such cases, the Monte Carlo approximation is often used to approximate the calculation.

We can now represent the high-level mathematical details of BO, ignoring the decision on termination for now. As shown in Figure 3-2, the whole BO (outer) loop consists of three major steps: proposing the following sampling location x_{n+1}^* as the maximizing location of the acquisition function $\alpha\left(x_{n+1}; \mathcal{D}_n\right)$ based on current dataset $\mathcal{D}_n$, probing the proposed location and appending the additional observation in the current

dataset $\mathcal{D}_{n+1} = \mathcal{D}_n \cup \left\{\left(x_{n+1}^*, y_{n+1}^*\right)\right\}$, and finally updating the posterior belief assuming a

GP surrogate model $p\left(f|\mathcal{D}_{n+1}\right)$. Here, y_{n+1}^* denotes the observation at the optimal next

sampling location x_{n+1}^*. Choosing an appropriate acquisition function thus plays a crucial role in determining the quality of the sequential optimization process.

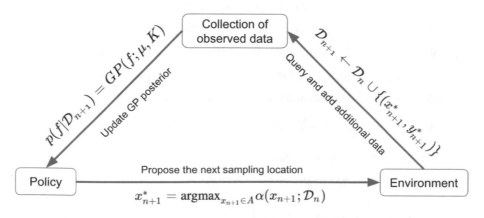

Figure 3-2. *Illustrating the entire BO loop by iteratively maximizing the current acquisition function, probing additional data, and updating posterior belief*

Although the BO loop could begin with an empty dataset, practical training often relies on a small dataset consisting of a few uniformly sampled observations. This accelerates the optimization process as it serves as a warm start and presents a more informed prior belief than a uniform one. The effect is even more evident when the initial dataset has good coverage of different locations of the domain.

Utility-Driven Optimization

The eventual goal of BO is to collect a valuable set of observations that are most informative about the global optimum. The value of a dataset is quantified by *utility*, a notion initially used in the Bayesian decision theory and used here to assist the sequential optimization in BO via the acquisition function. The acquisition function builds on top of the utility of the currently available dataset when assessing the value of candidate locations.

Since our goal is to locate the global maximum, a natural choice for the utility function is the maximum value of the current dataset, that is, $u(\mathcal{D}_n) = \max\{y_{1:n}\} = y_n^*$, assuming the case of noise-free observations. This is also called the incumbent of the current dataset and is used as a benchmark when evaluating all future candidate observations. As the most widely used acquisition function in practical applications, the expected improvement function uses this incumbent to award candidate locations whose putative observations are likely to be higher.

When assessing a candidate location x_{n+1}, we would require a fictional observation y to be able to calculate the utility *if* we were to acquire an additional observation at this location. Considering the randomness of the objective function, our best estimate is that y_{n+1} will follow a posterior normal distribution according to the updated GP posterior. Since y_{n+1} is a random variable, the standard approach is to integrate out its randomness by calculating the *expected utility* at the particular location, that is,

$\mathbb{E}_{y_{n+1}}\left[u(x_{n+1}, y_{n+1}, \mathcal{D}_n) \mid x_{n+1}, \mathcal{D}_n\right]$, conditioned on the specific evaluation location x_{n+1} and current set of observations $\mathcal{D}_n$. This also corresponds to the expected utility when assuming we have an additional unknown observation (x_{n+1}, y_{n+1}), leading to

$$\mathbb{E}_{y_{n+1}}\left[u(\mathcal{D}_{n+1}) \mid x_{n+1}, \mathcal{D}_n\right] = \mathbb{E}_{y_{n+1}}\left[u(\mathcal{D}_n \cup (x_{n+1}, y_{n+1})) \mid x_{n+1}, \mathcal{D}_n\right] = \mathbb{E}_{y_{n+1}}\left[u(x_{n+1}, y_{n+1}, \mathcal{D}_n) \mid x_{n+1}, \mathcal{D}_n\right]$$

. We could then utilize the posterior predictive distribution $p(y_{n+1} \mid \mathcal{D}_n)$ to express the expected utility as an integration operation in the continuous case as follows:

$$\mathbb{E}_{y_{n+1}}\left[u(x_{n+1}, y_{n+1}, \mathcal{D}_n) \mid x_{n+1}, \mathcal{D}_n\right] = \int u(x_{n+1}, y_{n+1}, \mathcal{D}_n) p(y_{n+1} \mid x_{n+1}, \mathcal{D}_n) dy_{n+1}$$

This expression considers all possible values of y_{n+1} at location x_{n+1}. It weighs the corresponding utility based on the probability of occurrence. With access to the expected utility at each candidate location, the next following location could be determined by selecting the one with the largest expected utility:

$$x_{n+1}^* = \text{argmax}_{x_{n+1} \in A} \mathbb{E}_{y_{n+1}} \left[u(x_{n+1}, y_{n+1}, \mathcal{D}_n) | x_{n+1}, \mathcal{D}_n \right]$$

Therefore, we need to have an appropriately designed utility function when determining the next optimal location by maximizing the expected utility. Equivalently, each action taken by the policy is selected to maximize the improvement in the expected utility. This process continues until the stopping rule is triggered, at which point the quality of the final returned dataset $\mathcal{D}_N$ is evaluated using $u(\mathcal{D}_N)$.

Since we are concerned with the optimal one-step lookahead action, the preceding problem can be formulated as maximizing the *expected marginal gain* in the utility, which serves as the acquisition function to guide the search. The one-step lookahead policy using the expected marginal gain is thus defined as follows:

$$\alpha_1(x_{n+1}; \mathcal{D}_n) = \mathbb{E}_{y_{n+1}} \left[u(\mathcal{D}_{n+1}) | x_{n+1}, \mathcal{D}_n \right] - \mathbb{E} \left[u(\mathcal{D}_n) \right]$$

$$= \mathbb{E}_{y_{n+1}} \left[u(x_{n+1}, y_{n+1}, \mathcal{D}_n) | x_{n+1}, \mathcal{D}_n \right] - u(\mathcal{D}_n)$$

where the subscript 1 in $\alpha_1(x_{n+1}; \mathcal{D}_n)$ denotes the number of lookahead steps into the future. The second step follows since there is no randomness in the utility of the existing observations $u(\mathcal{D}_n)$.

The optimal action using the one-step lookahead policy is then defined as the maximizer of the expected marginal gain:

$$x_{n+1}^* = \text{argmax}_{x_{n+1} \in A} \alpha_1(x_{n+1}; \mathcal{D}_n)$$

Figure 3-3 illustrates this process. We start with the utility of collected observations $u(\mathcal{D}_n)$ as the benchmark for comparison when evaluating the expected marginal gain at a new candidate location. The evaluation needs to consider all possible values of the next observation based on updated posterior GP and thus leads to the expected utility term $\mathbb{E}_{y_{n+1}} \left[u(x_{n+1}, y_{n+1}, \mathcal{D}_n) | x_{n+1}, \mathcal{D}_n \right]$. Since we are considering one step ahead in the future, the acquisition function $\alpha_1(x_{n+1}; \mathcal{D}_n)$ becomes one-step lookahead policy, and our goal is to select the location that maximizes the expected marginal gain in the utility of the collected dataset.

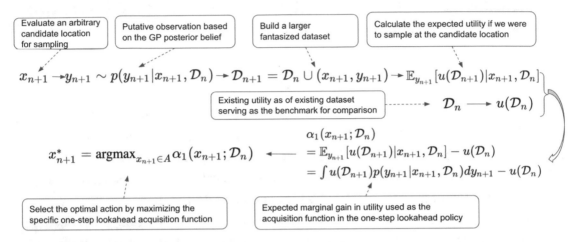

Figure 3-3. *Deriving the one-step lookahead policy by maximizing the expected marginal gain in the utility of the acquired observations*

Multi-step Lookahead Policy

The sequential decision-making process using a one-step lookahead policy is a powerful and widely applied technique. By simulating possible future paths if we were to collect another observation, the policy becomes Bayes optimal due to maximizing the one-step expected marginal gain of the utility in the enlarged artificial dataset. However, the optimization process will continue until reaching a terminal point when the search budget is exhausted. The choice of the following sampling location, $x_{n+1}^* = \mathrm{argmax}_{x_{n+1} \in A} \alpha_1(x_{n+1}; \mathcal{D}_n)$, thus impacts all remaining optimization decisions. That is, we need to consider all the future sampling steps until the stopping rule triggers, instead of only one step into the future.

To put things into context, let us assume that the *lookahead horizon*, that is, the number of steps to consider in the future, is τ. In other words, we would like to consider a putative dataset $\mathcal{D}_{n+\tau}$, which has additional τ artificial observations added to the existing dataset $\mathcal{D}_n$. Each observation involves selecting a candidate search location x and acquiring the corresponding observation value y, modeled as a random variable with updated posterior distribution based on previous observations (including both existing and putative ones). By expressing each addition of location and observation as a pair (x, y), the τ-step lookahead dataset $\mathcal{D}_{n+\tau}$ could be written as

$$\mathcal{D}_{n+\tau} = \mathcal{D}_n \cup \left\{ \left(x_{n+1}, y_{n+1} \right) \right\} \cup \left\{ \left(x_{n+2}, y_{n+2} \right) \right\} \cup \ldots \cup \left\{ \left(x_{n+\tau}, y_{n+\tau} \right) \right\}$$

Following the same mechanics as before, the multi-step lookahead policy would make the optimal sampling decision on x_{n+1}^* by maximizing the expected long-term terminal utility $\mathbb{E}\left[u\left(\mathcal{D}_{n+\tau} \right) \right]$:

$$x_{n+1}^* = \operatorname{argmax}_{x \in A} \mathbb{E}\left[u\left(\mathcal{D}_{n+\tau} \right) | x_{n+1}, \mathcal{D}_n \right]$$

where the expectation is taken with respect to randomness in future locations and observations. Equivalently, we can rely on the terminal expected marginal gain in the utility defined as follows:

$$\alpha_\tau \left(x_{n+1}; \mathcal{D}_n \right) = \mathbb{E}\left[u\left(\mathcal{D}_{n+\tau} \right) | x_{n+1}, \mathcal{D}_n \right] - u\left(\mathcal{D}_n \right)$$

which serves as the multi-step lookahead acquisition function to support the optimal sequential optimization:

$$x_{n+1}^* = \operatorname{argmax}_{x_{n+1} \in A} \alpha_\tau \left(x_{n+1}; \mathcal{D}_n \right)$$

where the definition is only shifted downward by a constant value $u\left(\mathcal{D}_n \right)$ compared with maximizing the expected terminal utility $\mathbb{E}\left[u\left(\mathcal{D}_{n+\tau} \right) | x, \mathcal{D}_n \right]$ alone.

Now, if we expand the expectation in the definition of $\alpha_\tau \left(x_{n+1}; \mathcal{D}_n \right)$, we would need to consider all possible evolutions of future τ-step decisions on the locations $\{x_{n+i}, i = 2, \ldots, \tau\}$ and the associated realizations of the random variables $\{y_{n+i}, i = 1, \ldots, \tau\}$. Here, decisions on the locations $\{x_{n+i}, i = 2, \ldots, \tau\}$ start with $i = 2$ due to the fact that we are evaluating at location x_{n+1}. We can write the expanded form of the terminal expected marginal gain in utility as follows:

$$\alpha_\tau \left(x_{n+1}; \mathcal{D}_n \right) = \int \cdots \int u\left(\mathcal{D}_{n+\tau} \right) p\left(y_{n+1} | x_{n+1}, \mathcal{D}_n \right) \prod_{i=2}^{\tau} p\left(x_{n+i}, y_{n+i} | \mathcal{D}_{n+i-1} \right) dy_{n+1} d\left\{ \left(x_{n+i}, y_{n+i} \right) \right\} - u\left(\mathcal{D}_n \right)$$

where we explicitly write the posterior probability distribution of y_{n+1} as $p\left(y_{n+1} | x_{n+1}, \mathcal{D}_n \right)$ and the following joint probability distributions of $\{(x_{n+i}, y_{n+i}), i = 2, \ldots, \tau\}$ as

$\prod_{i=2}^{\tau} p\left(x_{n+i}, y_{n+i} | \mathcal{D}_{n+i-1} \right)$. Integrating out these random variables would give us the eventual multi-step lookahead marginal gain in the expected utility of the returned dataset.

Figure 3-4 summarizes the process of deriving the multi-step lookahead acquisition function. Note that the simulation of the next round of candidate locations and observations in $\{(x_{n+i}, y_{n+i}), i = 2, ..., \tau\}$ depends on all previously accumulated dataset $\mathcal{D}_{n+i-1}$, which is used to construct the updated posterior belief based on both observed and putative values.

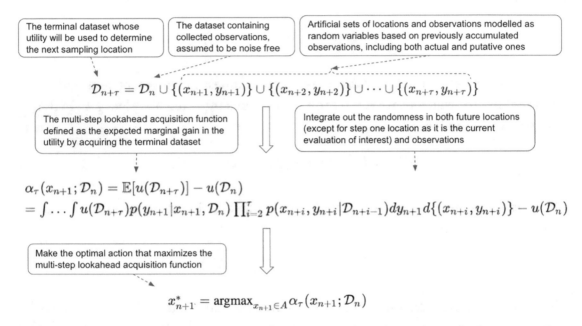

Figure 3-4. *The multi-step lookahead optimal policy that selects the best sampling location by maximizing the marginal expected utility of the terminal dataset*

We can glean more insight on the process of calculating this expression by drawing out the sequence of nested expectation and maximization operations. As shown in Figure 3-5, we start with the next sampling location x_{n+1} in a maximization operator, followed by y_{n+1} in an expectation operator. The same pattern continues at later stages, with a maximization operator in x_{n+2}, an expectation operator in y_{n+2}, and so on, until reaching the putative observation $y_{n+\tau}$ at the last stage. Each operator, be it maximization or expectation, involves multiple branches. Common strategy is to solve the maximization operation via a standard procedure such as L-BFGS and approximate the expectation operation via Gaussian quadrature.

$$x_{n+1}^* = \text{argmax}_{x_{n+1}\in X} \alpha_\tau(x_{n+1}; D_n)$$

$$= \int \dots \int u(D_{n+\tau}) p(y_{n+1}|x_{n+1}, D_n) \prod_{i=2}^{\tau} p(x_{n+i}, y_{n+i}|D_{n+i-1}) \, dy_{n+1} d\{(x_{n+i}, y_{n+i})\} - u(D_n)$$

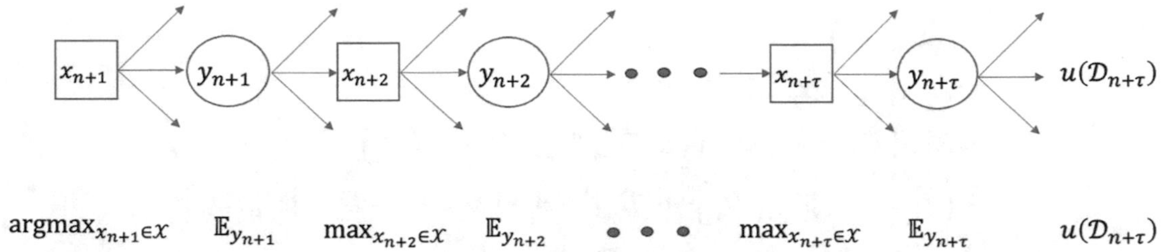

$$\text{argmax}_{x_{n+1}\in X} \quad \mathbb{E}_{y_{n+1}} \quad \text{max}_{x_{n+2}\in X} \quad \mathbb{E}_{y_{n+2}} \quad \bullet \bullet \bullet \quad \text{max}_{x_{n+\tau}\in X} \quad \mathbb{E}_{y_{n+\tau}} \quad u(D_{n+\tau})$$

Figure 3-5. *Visualizing the nested maximization and expectation operators*

Apparently, calculating a nested form of expectations that accounts for all possible future paths is computationally challenging. In addition, since our goal is to select an optimal sampling action by maximizing the acquisition function, we will add a reasonable assumption that all future actions will also be optimal given the current dataset, which may include putative realizations of the random variable on the objective value. Adding the optimality condition means that rather than considering all possible future paths of $\{(x_{n+i}, y_{n+i}), i = 1, \dots, \tau\}$, we will only focus on the optimal one $\left\{\left(x_{n+i}^*, y_{n+i}\right), i = 1, \dots, \tau\right\}$, which essentially removes the dependence on the candidate locations by choosing the maximizing location. The argument for selecting the optimal action by maximizing the long-term expected gain in utility follows the *Bellman principle of optimality*, as described in the next section.

Bellman's Principle of Optimality

Bellman's principle of optimality states that a sequence of optimal decisions starts with making the first optimal decision, followed by a series of optimal decisions conditioned on the previous outcome. This is a recursive expression in that, in order to make an optimal action at the current time point, we will need to act optimally in the future.

Let us build from the multi-step lookahead acquisition function from earlier. Recall that the τ-step expected marginal gain in utility at a candidate location x_{n+1} is defined as

$$\alpha_\tau\left(x_{n+1}; D_n\right) = \mathbb{E}\left[u\left(D_{n+\tau}\right)|x_{n+1}, D_n\right] - u\left(D_n\right)$$

which is the subject we seek to maximize. To explicitly connect with the one-step lookahead acquisition function and the remaining $\tau - 1$ steps of simulations into the future, we can introduce the one-step utility $u(\mathcal{D}_{n+1})$ by adding and subtracting this term in the expectation, as shown in the following:

$$
\begin{aligned}
&\alpha_\tau\left(x_{n+1};\mathcal{D}_n\right)\\
&=\mathbb{E}\left[u\left(\mathcal{D}_{n+\tau}\right)|x_{n+1},\mathcal{D}_n\right]-u\left(\mathcal{D}_n\right)\\
&=\mathbb{E}\left[u\left(\mathcal{D}_{n+\tau}\right)-u\left(\mathcal{D}_{n+1}\right)+u\left(\mathcal{D}_{n+1}\right)|x_{n+1},\mathcal{D}_n\right]-u\left(\mathcal{D}_n\right)\\
&=\left(\mathbb{E}\left[u\left(\mathcal{D}_{n+1}\right)|x_{n+1},\mathcal{D}_n\right]-u\left(\mathcal{D}_n\right)\right)+\mathbb{E}\left[u\left(\mathcal{D}_{n+\tau}\right)-u\left(\mathcal{D}_{n+1}\right)|x_{n+1},\mathcal{D}_n\right]\\
&=\alpha_1\left(x_{n+1};\mathcal{D}_n\right)+\mathbb{E}\left[\alpha_{\tau-1}\left(x_{n+2};\mathcal{D}_{n+1}\right)|x_{n+1},\mathcal{D}_n\right]
\end{aligned}
$$

Here, we have decomposed the long-term expected marginal gain in utility into the sum of an immediate one-step lookahead gain in utility and the expected lookahead gain for the remaining $\tau - 1$ steps.

Now, following Bellman's principle of optimality, all the remaining $\tau - 1$ actions will be made optimally. This means that instead of evaluating each candidate location for x_{n+2} when calculating $\alpha_{\tau-1}\left(x_{n+2};\mathcal{D}_{n+1}\right)$, we would only be interested in the location with the maximal value, that is, $\alpha_{\tau-1}\left(x_{n+2}^*;\mathcal{D}_{n+1}\right)$, or equivalently $\alpha_{\tau-1}^*\left(\mathcal{D}_{n+1}\right)$, removing dependence on the location x_{n+2}. The multi-step lookahead acquisition function under the optimality assumption thus becomes

$$
\alpha_\tau\left(x_{n+1};\mathcal{D}_n\right)=\alpha_1\left(x_{n+1};\mathcal{D}_n\right)+\mathbb{E}\left[\alpha_{\tau-1}^*\left(\mathcal{D}_{n+1}\right)|x_{n+1},\mathcal{D}_n\right]
$$

As shown in the previous section, the optimal next sampling location x_{n+1}^* using the multi-step lookahead acquisition function is thus determined by maximizing $\alpha_\tau\left(x_{n+1};\mathcal{D}_n\right)$. The optimal multi-step lookahead acquisition function $\alpha_\tau^*\left(x_{n+1};\mathcal{D}_n\right)$ is thus defined as

$$
\alpha_\tau^*\left(x_{n+1};\mathcal{D}_n\right)=\max_{x_{n+1}\in A}\alpha_\tau\left(x_{n+1};\mathcal{D}_n\right)
$$

Plugging in the definition of $\alpha_\tau\left(x_{n+1};\mathcal{D}_n\right)$ gives

$$
\begin{aligned}
&\alpha_\tau^*\left(x_{n+1};\mathcal{D}_n\right)\\
&=\max_{x_{n+1}\in A}\left\{\alpha_1\left(x_{n+1};\mathcal{D}_n\right)+\mathbb{E}\left[\alpha_{\tau-1}^*\left(\mathcal{D}_{n+1}\right)|x_{n+1},\mathcal{D}_n\right]\right\}\\
&=\max_{x_{n+1}\in A}\left\{\alpha_1\left(x_{n+1};\mathcal{D}_n\right)+\mathbb{E}\left[\max_{x_{n+2}\in A}\alpha_{\tau-1}\left(x_{n+2};\mathcal{D}_{n+1}\right)|x_{n+1},\mathcal{D}_n\right]\right\}
\end{aligned}
$$

where we have plugged in the definition of $\alpha_{\tau-1}^*(\mathcal{D}_{n+1})$ as well to explicitly express the optimal policy value $\alpha_\tau^*(x_{n+1};\mathcal{D}_n)$ as a series of nested maximization and expectation operations. Such recursive definition is called the *Bellman equation*, which explicitly reflects the condition that all follow-up actions need to be made optimally to make an optimal action.

Figure 3-6 summarizes the process of deriving the Bellman equation for the multi-step lookahead policy. Again, calculating the optimal policy value requires calculating the expected optimal value of future subpolicies. Being recursive in nature, calculating the current acquisition function can be achieved by adopting a reverse computation, starting from the terminal step and performing the calculations backward. However, this would still incur an exponentially increasing burden as the lookahead horizon expands.

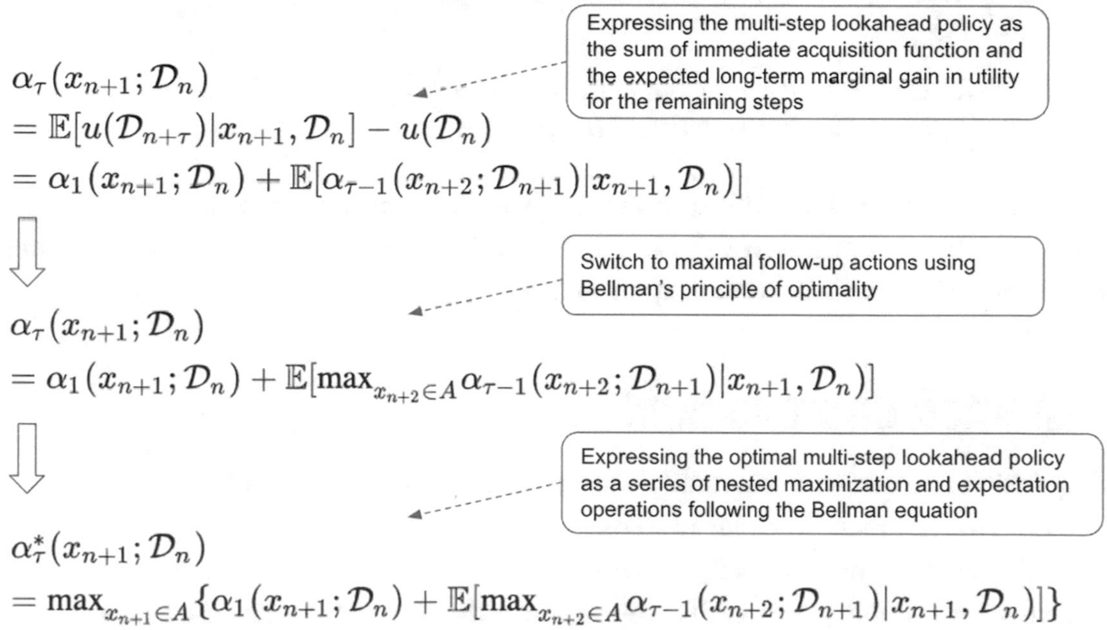

$\alpha_\tau(x_{n+1};\mathcal{D}_n)$

$= \mathbb{E}[u(\mathcal{D}_{n+\tau})|x_{n+1},\mathcal{D}_n] - u(\mathcal{D}_n)$

$= \alpha_1(x_{n+1};\mathcal{D}_n) + \mathbb{E}[\alpha_{\tau-1}(x_{n+2};\mathcal{D}_{n+1})|x_{n+1},\mathcal{D}_n)]$

> Expressing the multi-step lookahead policy as the sum of immediate acquisition function and the expected long-term marginal gain in utility for the remaining steps

$\Downarrow$

> Switch to maximal follow-up actions using Bellman's principle of optimality

$\alpha_\tau(x_{n+1};\mathcal{D}_n)$

$= \alpha_1(x_{n+1};\mathcal{D}_n) + \mathbb{E}[\max_{x_{n+2}\in A}\alpha_{\tau-1}(x_{n+2};\mathcal{D}_{n+1})|x_{n+1},\mathcal{D}_n)]$

$\Downarrow$

> Expressing the optimal multi-step lookahead policy as a series of nested maximization and expectation operations following the Bellman equation

$\alpha_\tau^*(x_{n+1};\mathcal{D}_n)$

$= \max_{x_{n+1}\in A}\{\alpha_1(x_{n+1};\mathcal{D}_n) + \mathbb{E}[\max_{x_{n+2}\in A}\alpha_{\tau-1}(x_{n+2};\mathcal{D}_{n+1})|x_{n+1},\mathcal{D}_n)]\}$

Figure 3-6. *Illustrating the derivation process of the Bellman equation for the multi-step lookahead policy, where the optimal policy is expressed as a series of maximization and expectation operations, assuming all follow-up actions need to be made optimally in order to make the optimal action at the current step*

We will touch upon several tricks to accelerate the calculation of this *dynamic programming* (DP) problem later in the book and only highlight two common approaches for now. One approach is called limited lookahead, which limits the number of lookahead steps in the future. The other is to use a rollout approach with a base policy, which reduces the maximization operator into a quick heuristic-based exercise. Both approaches are called approximate dynamic programming (ADP) methods and are illustrated in Figure 3-7. See the recent book titled *Bayesian Optimization* by Roman Garnett for more discussion on this topic.

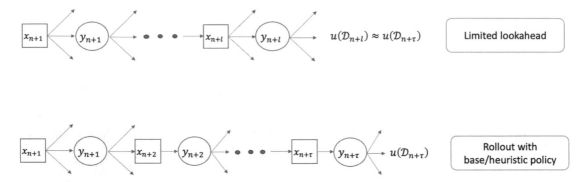

Figure 3-7. *Two approximate dynamic programming approaches commonly used to calculate the multi-step lookahead BO policies*

In the next section, we will introduce the expected improvement acquisition function, which is the most widely used and empirically performing acquisition function in practical Bayesian optimization applications.

Expected Improvement

Acquisition functions differ in multiple aspects, including the choice of the utility function, the number of lookahead steps, the level of risk aversion or preference, etc. Introducing risk appetite directly benefits from the posterior belief about the underlying objective function. In the case of GP regression as the surrogate model, the risk is quantified by the covariance function, with its credible interval expressing the level of uncertainty about the possible values of the objective.

When it comes to the utility of the collected observations, the expected improvement chooses the maximum of the observed value as the benchmark for comparison upon selecting an additional sampling location. It also implicitly assumes that only one sampling is left before the optimization process terminates. The expected marginal gain in utility (i.e., the acquisition function) becomes the expected improvement in the maximal observation, calculated as the expected difference between the observed maximum and the new observation after the additional sampling at an arbitrary sampling location.

These assumptions make the expected improvement a one-step lookahead acquisition function, also called *myopic* due to its short lookahead horizon. Besides, since the expectation of the posterior distribution is used, the expected improvement is also considered *risk neutral*, disregarding the uncertainty estimates across the whole domain.

Specifically, denote $y_{1:n} = \{y_1, ..., y_n\}$ as the set of collected observations at the corresponding locations $x_{1:n} = \{x_1, ..., x_n\}$. Assuming the noise-free setting, the actual observations are exact, that is, $y_{1:n} = f_{1:n}$. Given the collected dataset $\mathcal{D}_n = \{x_{1:n}, y_{1:n}\}$, the corresponding utility is $u(\mathcal{D}_n) = \max\{f_{1:n}\} = f_n^*$, where f_n^* is the incumbent maximum observed so far. Similarly, assuming we obtain another observation $y_{n+1} = f_{n+1}$ at a new location x_{n+1}, the resulting utility is $u(\mathcal{D}_{n+1}) = u(\mathcal{D}_n \cup \{x_{n+1}, f_{n+1}\}) = \max\{f_{n+1}, f_n^*\}$. Taking the difference of these two gives the increase in utility due to the addition of another observation:

$$u(\mathcal{D}_{n+1}) - u(\mathcal{D}_n) = \max\{f_{n+1}, f_n^*\} - f_n^* = \max\{f_{n+1} - f_n^*, 0\}$$

which returns the marginal increment in the incumbent if $f_{n+1} \geq f_n^*$ and zero otherwise, as a result of observing f_{n+1}. Readers familiar with the activation function in neural networks would instantly connect this form with the ReLU (rectified linear unit) function, which keeps the positive signal as it is and silences the negative one.

Due to randomness in y_{n+1}, we can introduce the expectation operator to integrate it out, giving us the expected marginal gain in utility, that is, the expected improvement acquisition function:

$$\alpha_{\mathrm{EI}}(x_{n+1}; \mathcal{D}_n) = \mathbb{E}\left[u(\mathcal{D}_{n+1}) - u(\mathcal{D}_n)|x_{n+1}, \mathcal{D}_n\right]$$
$$= \int \max\{f_{n+1} - f_n^*, 0\} p(f_{n+1}|x_{n+1}, \mathcal{D}_n) df_{n+1}$$

Under the framework of GP regression, we can obtain a closed-form expression of the expected improvement acquisition function, as shown in the following section.

Deriving the Closed-Form Expression

The expected improvement acquisition function admits a convenient closed-form expression, which could significantly accelerate its computation. Deriving the closed-form expression requires the scale-location transformation from a standard normal variable to an arbitrary normally distributed variable, as covered in a previous chapter. This is also called the *reparameterization trick* since we can convert the subject of interest into a standard normal variable to simplify mathematical analysis and practical computation.

Precisely, since the observation f_{n+1} at candidate location x_{n+1} follows a normal distribution with the corresponding posterior mean μ_{n+1} and variance σ_{n+1}^2, writing $f_{n+1} \sim N\left(f_{n+1}; \mu_{n+1}, \sigma_{n+1}^2\right)$, we can reparameterize it as $f_{n+1} = \mu_{n+1} + \sigma_{n+1}\varepsilon$, where $\varepsilon \sim N(\varepsilon; 0, 1)$.

Figure 3-8 gives the full derivation process that involves a few technical details such as linearity of expectation, integration by parts, the standard and cumulative standard normal distribution, and change of variable in differentiation. Assuming $\sigma_{n+1}^2 > 0$, the process starts by converting the max operator into an integral, which is then separated into two different and easily computable parts. These two parts correspond to exploitation and exploration, respectively. Exploitation means continuing sampling the neighborhood of the observed region with a high posterior mean, and exploration encourages sampling an unvisited area where the posterior uncertainty is high. The expected improvement acquisition function thus implicitly balances off these two opposing forces.

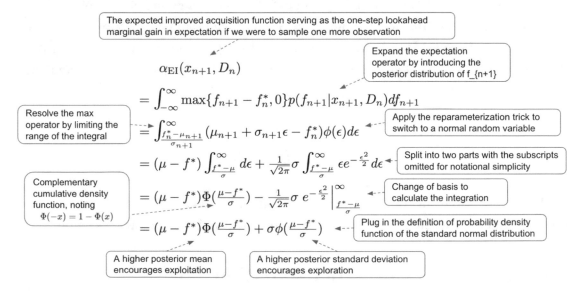

Figure 3-8. *Deriving the closed-form expression of expected improvement, which automatically balances between the exploitation of promising areas given existing knowledge and exploration of uncertain areas*

To further establish the monotonic relationship between the posterior parameters $(\mu_{n+1}$ and $\sigma_{n+1}^2)$ and the value of the expected improvement, we could examine the respective partial derivative. Concretely, we have the following:

$$\frac{\partial}{\partial \mu_{n+1}} \alpha_{\mathrm{EI}}\left(x_{n+1}; \mathcal{D}_n\right) = \Phi\left(\frac{\mu_{n+1} - f^*}{\sigma_{n+1}}\right) > 0$$

$$\frac{\partial}{\partial \sigma_{n+1}} \alpha_{\mathrm{EI}}\left(x_{n+1}; \mathcal{D}_n\right) = \phi\left(\frac{\mu_{n+1} - f^*}{\sigma_{n+1}}\right) > 0$$

Since the partial derivatives of the expected improvement with respect to μ_{n+1} and σ_{n+1} are both positive, an increase in either parameter will result in a higher expected improvement, thus completing the automatic trade-off between exploitation and exploration under the GP regression framework.

It is also worth noting that $\sigma_{n+1} = 0$ occurs when the posterior mean function passes through the observations. In this case, we have $\alpha_{\mathrm{EI}}\left(x_{n+1}; \mathcal{D}_n\right) = 0$. In addition, a hyperparameter ξ is often introduced to control the amount of exploration in practical implementation. By subtracting ξ from $\mu_{n+1} - f_n^*$ in the preceding closed-form expression, the posterior mean μ_{n+1} will have less impact on the overall improvement compared to the posterior standard deviation σ_{n+1}. The closed-form expression of the expected improvement acquisition function thus becomes

$$\alpha_{\mathrm{EI}}\left(x_{n+1}; \mathcal{D}_n\right) = \begin{cases} \left(\mu_{n+1} - f_n^* - \xi\right)\Phi\left(z_{n+1}\right) + \sigma_{n+1}\phi\left(z_{n+1}\right); & \sigma_{n+1} > 0 \\ 0; & \sigma_{n+1} = 0 \end{cases}$$

where

$$z_{n+1} = \begin{cases} \dfrac{\mu_{n+1} - f^* - \xi}{\sigma_{n+1}}; & \sigma_{n+1} > 0 \\ 0; & \sigma_{n+1} = 0 \end{cases}$$

The following section will implement the expected improvement acquisition function and use it to look for the global optimum of synthetic test functions.

Implementing the Expected Improvement

In this section, we will first implement the expected improvement acquisition function from scratch based on plain NumPy and SciPy packages, followed by using an off-the-shelf BO package to complete the same task. The true underlying objective function is given to us for comparison and unrevealed to the optimizer, whose goal is to approximate the objective function from potentially noise samples. The codes are adapted from a tutorial blog from Martin Krasser (`http://krasserm.github.io/2018/03/21/bayesian-optimization/#Optimization-algorithm`).

First, we will set up the environment by importing a few packages for Gaussian process regression from scikit-learn, numerical optimization from SciPy, and other utility functions on plotting. We also set the random seed to ensure reproducibility.

Listing 3-1. Setting up the coding environment

```
import numpy as np
import random
import matplotlib.pyplot as plt
from scipy.stats import norm
from scipy.optimize import minimize
from sklearn.gaussian_process import GaussianProcessRegressor
from sklearn.gaussian_process.kernels import ConstantKernel, Matern

SEED = 8
random.seed(SEED)
np.random.seed(SEED)
%matplotlib inline
```

Next, we will define the objective function and search domain. The objective function provides noise-perturbed observations upon sampling at an arbitrary location within the search domain. It will also be used to generate noise-free observations for reference during plotting.

As shown in the following code listing, we generate a random number from a standard normal distribution based on the dimension of the domain, accessed via the * sign to unpack the tuple into an acceptable format. The value is then multiplied by the prespecified noise level for the observation model. The search domain is specified as a nested list in bounds, where the inner list contains the upper and lower bounds for each dimension; in this case, we are looking at a single-dimensional search domain.

Listing 3-2. Defining the search domain and objective function

```
# search bounds of the domain
# each element in the inner list corresponds to one dimension
bounds = np.array([[-1.0, 2.0]])
# observation noise
noise = 0.2
# objective function used to reveal observations upon sampling, optionally
with noise
def f(x, noise=0):
    # use * to unpack the tuple from x.shape when passing into
    np.random.randn
    return -np.sin(3*x) - x**2 + 0.7*x + noise*np.random.randn(*x.shape)
```

Now we can visualize the objective function and generate two random noisy samples in X_init and Y_init to kick-start the optimization procedure. Note that plotting a function is completed by generating a dense grid of points/locations with the search bounds in X_plot, calculating the corresponding noise-free functional values Y_plot for each location, and connecting these values smoothly, as shown in the following code listing.

Listing 3-3. Visualizing the underlying objective function and initial noisy samples

```
# initial observations upon initiation
X_init = np.array([[-0.7], [1.6]])
Y_init = f(X_init, noise=noise)
# dense grid of points within bounds used for plotting
X_plot = np.arange(bounds[:, 0], bounds[:, 1], 0.01).reshape(-1, 1)
# noise-free objective function values used for plotting
Y_plot = f(X_plot, noise=0)
# Plot objective function with noisy observations
plt.plot(X_plot, Y_plot, 'y--', lw=2, label='Objective function')
plt.plot(X_init, Y_init, 'kx', mew=3, label='Initial noisy samples')
plt.legend()
```

The result is shown in Figure 3-9. Note that the two samples are selected to be sufficiently distant from each other. In practice, a good initial design should have good coverage of the whole search domain to promise a good GP prior before optimization starts.

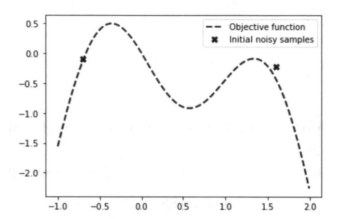

Figure 3-9. *Visualizing the underlying objective function and two initial random noisy samples*

We now define the expected improvement acquisition function as our sampling policy. This function maps each sampling location input to a numeric scalar output, the expected marginal gain in utility. In the following code listing, other than the evaluation points in X, the inputs also include the previously observed locations X_sample and values Y_sample, along with a GP regressor gpr fitted to the training samples. Besides, we also include the hyperparameter xi to control the level of exploration with a default value of 0.01.

Listing 3-4. Defining the expected improvement acquisition function

```
def expected_improvement(X, X_sample, Y_sample, gpr, xi=0.01):
    # posterior mean and sd at proposed location
    mu, sigma = gpr.predict(X, return_std=True)
    # posterior mean at observed location
    mu_sample = gpr.predict(X_sample)
    # reshape to make one sd per each proposed location
    sigma = sigma.reshape(-1, 1)
    # use maximal posterior mean instead of actual observations due
    to noise
```

```
mu_sample_opt = np.max(mu_sample)
# ignore divide by zero warning if any
with np.errstate(divide='warn'):
    # calculate ei if sd>0
    imp = mu - mu_sample_opt - xi
    Z = imp / sigma
    ei = imp * norm.cdf(Z) + sigma * norm.pdf(Z)
    # set zero if sd=0
    ei[sigma == 0.0] = 0.0
return ei
```

Note that we start by plugging in the definition of expected improvement assuming a nonzero standard deviation for the posterior distribution at the proposed location, followed by setting the entries with zero standard deviation to be zero. Since directly dividing by zero gives an error, as needed when calculating Z = imp / sigma, the calculation is moved within the context of np.errstate(divide='warn'), which is a particular arrangement to tell Python to temporarily ignore such error because of the follow-up treatment via ei[sigma == 0.0] = 0.0.

At this stage, we can calculate the expected improvement of any candidate location, and our goal is to find the optimal location with the biggest value in expected improvement. To achieve this, we will use a particular off-the-shelf optimizer called "L-BFGS-B" provided by the minimize function from SciPy, which utilizes the approximate second-order derivative to solve for the optimum of a given function, that is, the expected improvement. The location of the optimum can be retrieved at the end of the optimization procedure.

The following code listing defines a function called propose_location() that performs optimization for a total of n_restarts rounds so as to avoid local optima. By keeping a running minimum min_val and its location min_x, each round of optimization returns an optimal solution via minimizing the negative of the acquisition function value via the min_obj() function; maximizing a positive value is equivalent to minimizing its negative. At last, we decide if the current running minimum and the location need to be replaced by comparing it with the optimization solution. This function also completes the *inner loop* of BO, as introduced earlier.

Listing 3-5. Proposing the next sampling point by optimizing the acquisition function

```python
def propose_location(acquisition, X_sample, Y_sample, gpr, bounds,
n_restarts=25):
    # dimension of search domain
    dim = X_sample.shape[1]
    # temporary running best minimum
    min_val = 1
    # temporary location of best minimum
    min_x = None

    # map an arbitrary location to the negative of acquisition function
    def min_obj(X):
        # Minimization objective is the negative acquisition function
        return -acquisition(X.reshape(-1, dim), X_sample, Y_sample, gpr)

    # iterate through n_restart different random points and return most
    promising result
    for x0 in np.random.uniform(bounds[:, 0], bounds[:, 1],
    size=(n_restarts, dim)):
        # use off-the-shelf solver based on approximate second order
        derivative
        res = minimize(min_obj, x0=x0, bounds=bounds, method='L-BFGS-B')
        # replace running optimum if any
        if res.fun < min_val:
            min_val = res.fun[0]
            min_x = res.x

    return min_x.reshape(-1, 1)
```

Before entering BO's outer loop to seek the global optimum, we will define a few utility functions that plot the policy performance across iterations. This includes the `plot_approximation()` function that plots the GP posterior mean and 95% confidence interval along with the collected samples and objective function, the `plot_acquisition()` function that plots the expected improvement across the domain along with the location of the maximum, and the plot_convergence() function that plots the distances between consecutive sampling locations and the running optimal value as optimization proceeds. All three functions are defined in the following code listing.

Listing 3-6. Proposing the next sampling point by optimizing the acquisition function

```python
def plot_approximation(gpr, X_plot, Y_plot, X_sample, Y_sample,
X_next=None, show_legend=False):
    # get posterior mean and sd across teh dense grid
    mu, std = gpr.predict(X_plot, return_std=True)
    # plot mean and 95% confidence interval
    plt.fill_between(X_plot.ravel(),
                     mu.ravel() + 1.96 * std,
                     mu.ravel() - 1.96 * std,
                     alpha=0.1)
    plt.plot(X_plot, Y_plot, 'y--', lw=1, label='Noise-free objective')
    plt.plot(X_plot, mu, 'b-', lw=1, label='Surrogate function')
    plt.plot(X_sample, Y_sample, 'kx', mew=3, label='Noisy samples')
    # plot the next sampling location as vertical line
    if X_next:
        plt.axvline(x=X_next, ls='--', c='k', lw=1)
    if show_legend:
        plt.legend()

def plot_acquisition(X_plot, acq_value, X_next, show_legend=False):
    # plot the value of acquisition function across the dense grid
    plt.plot(X_plot, acq_value, 'r-', lw=1, label='Acquisition function')
    # plot the next sampling location as vertical line
    plt.axvline(x=X_next, ls='--', c='k', lw=1, label='Next sampling
    location')
    if show_legend:
        plt.legend()

def plot_convergence(X_sample, Y_sample, n_init=2):
    plt.figure(figsize=(12, 3))
    # focus on sampled queried by the optimization policy
```

```
x = X_sample[n_init:].ravel()
y = Y_sample[n_init:].ravel()
r = range(1, len(x)+1)
# distance between consecutive sampling locations
x_neighbor_dist = [np.abs(a-b) for a, b in zip(x, x[1:])]
# best observed value until the current time point
y_max = np.maximum.accumulate(y)
# plot the distance between consecutive sampling locations
plt.subplot(1, 2, 1)
plt.plot(r[1:], x_neighbor_dist, 'bo-')
plt.xlabel('Iteration')
plt.ylabel('Distance')
plt.title('Distance between consecutive x\'s')
# plot the evolution of observed maximum so far
plt.subplot(1, 2, 2)
plt.plot(r, y_max, 'ro-')
plt.xlabel('Iteration')
plt.ylabel('Best Y')
plt.title('Value of best selected sample')
```

Now we can move into the main outer loop to look for the global optimum by maximizing the expected improvement at each stage. In the following code listing, we first instantiate a GP regressor with a Matérn kernel, which accepts two hyperparameters that can be estimated by maximizing the marginal likelihood of the observed samples. In this case, we fix these hyperparameters to simplify the process. The GP regressor also accepts the unknown noise level via the alpha argument to incorporate noise in the observations.

Listing 3-7. The main BO loop

```
# Gaussian process with Matern kernel as surrogate model
# kernel parameters could be optimized using MLE
m52 = ConstantKernel(1.0) * Matern(length_scale=1.0, nu=2.5)
# specify observation noise term, assumed to be known in advance
gpr = GaussianProcessRegressor(kernel=m52, alpha=noise**2)
# initial samples before optimization starts
```

```python
X_sample = X_init
Y_sample = Y_init
# number of optimization iterations
n_iter = 20
# specify figure size
plt.figure(figsize=(12, n_iter * 3))
plt.subplots_adjust(hspace=0.4)
# start of optimization
for i in range(n_iter):
    # update GP posterior given existing samples
    gpr.fit(X_sample, Y_sample)
    # obtain next sampling point from the acquisition function (expected_
    improvement)
    X_next = propose_location(expected_improvement, X_sample, Y_sample,
    gpr, bounds)
    # obtain next noisy sample from the objective function
    Y_next = f(X_next, noise)
    # plot samples, surrogate function, noise-free objective and next
    sampling location
    plt.subplot(n_iter, 2, 2 * i + 1)
    plot_approximation(gpr, X_plot, Y_plot, X_sample, Y_sample, X_next,
    show_legend=i==0)
    plt.title(f'Iteration {i+1}')
    plt.subplot(n_iter, 2, 2 * i + 2)
    plot_acquisition(X_plot, expected_improvement(X_plot, X_sample, Y_
    sample, gpr), X_next, show_legend=i==0)
    # append the additional sample to previous samples
    X_sample = np.vstack((X_sample, X_next))
    Y_sample = np.vstack((Y_sample, Y_next))
```

Here, we use X_sample and Y_sample to be the running dataset augmented with additional samples as optimization continues for a total of 20 iterations. Each iteration consists of updating the GP posterior, locating the maximal expected improvement, observing at the proposed location, and incorporating the additional sample to the training set.

The codes also generate plots using `plot_approximation()` and `plot_acquisition()` to show more details on the optimization in each iteration. Figure 3-10 shows the first three iterations, where the optimizer exhibits an exploratory attribute by proposing samples relatively distant from each other. In other words, regions with high uncertainty are encouraged at the initial stage of optimization using the expected improvement acquisition function.

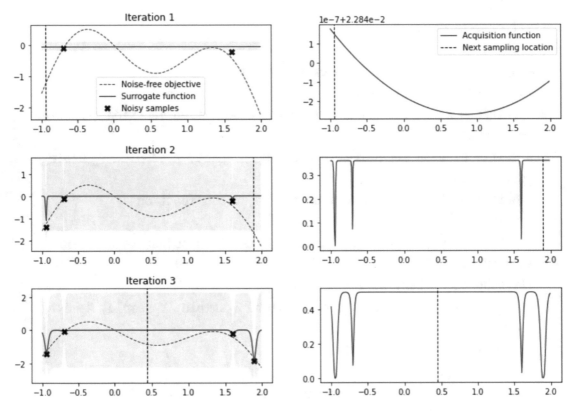

Figure 3-10. *Plotting the first three iterations, in which the EI-based BO performs more exploration at regions with high uncertainty*

As the optimization proceeds, the optimizer gradually resolves the uncertainty at distant locations and starts to rely more on exploitation of promising regions based on existing knowledge. This is reflected by the concentration of sampling locations at the left peak of the objective function, as shown by the last three iterations in Figure 3-11. Given that the last three sampling proposals occur at similar locations, we can roughly sense that the optimization process has converged, and the task of locating the global maximum is successfully completed.

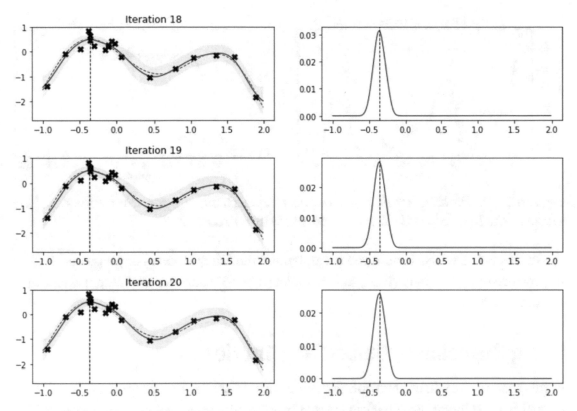

Figure 3-11. *Concentration of sampling locations at the left peak of the objective function, a sign of exploitation as the optimization process converges*

For the full list of intermediate plots across iterations, please visit the accompanying notebook for this chapter at `https://github.com/jackliu333/bayesian_optimization_theory_and_practice_with_python/blob/main/Chapter_3.ipynb`.

Once the optimization completes, we can examine its convergence using the `plot_convergence()` function. As shown in the left plot in Figure 3-12, a larger distance corresponds to more exploration, which occurs mostly at the initial stage of optimization as well as iterations 17 and 18 even when the optimization seems to be converging. Such exploration nature is automatically enabled by expected improvement and helps jumping out of local optima in search of a potentially higher global optimum. This is also reflected in the right plot, where a higher value is obtained at iteration 17 due to exploration.

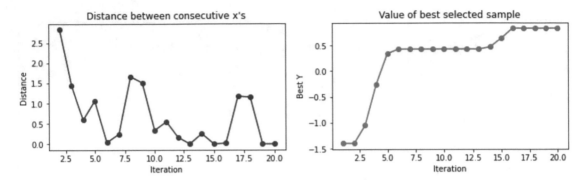

Figure 3-12. *Plotting the distance between consecutive proposed locations and the value of the best-selected sample as optimization proceeds*

At this point, we have managed to implement the full BO loop using expected improvement from scratch. Next, we will look at a few BO libraries that help us achieve the same task.

Using Bayesian Optimization Libraries

In this section, we will use two public Python-based libraries that support BO: `scikit-optimize` and `GPyOpt`. Both packages provide utility functions that perform BO after specifying the relevant input arguments. Let us look at optimizing the same function as earlier using `gp_minimize`, a function from `scikit-optimize` used to perform BO using GP.

In the following code listing, we specify the same kernel and hyperparameters setting for the GP instance `gpr`, along with the function `f` that provides noisy samples, search bounds `dimensions`, acquisition function `acq_func`, initial samples, exploration and exploitation trade-off parameter `xi`, number of iterations `n_calls`, as well as initial samples in `x0` and `y0`. At the end of optimization, we show the approximation plot to observe the locations of the proposed samples.

Listing 3-8. BO using scikit-optimize

```
from sklearn.base import clone
from skopt import gp_minimize
from skopt.learning import GaussianProcessRegressor
from skopt.learning.gaussian_process.kernels import ConstantKernel, Matern
```

```
# use custom kernel and estimator to match previous example
m52 = ConstantKernel(1.0) * Matern(length_scale=1.0, nu=2.5)
g = GaussianProcessRegressor(kernel=m52, alpha=noise**2)
# start BO
r = gp_minimize(func=lambda x: -f(np.array(x), noise=noise)[0], # function
to minimize
                dimensions=bounds.tolist(),    # search bounds
                base_estimator=gpr, # GP prior
                acq_func='EI',       # expected improvement
                xi=0.01,             # exploitation-exploration trade-off
                n_calls=n_iter,      # number of iterations
                n_initial_points=0, # initial samples are provided
                x0=X_init.tolist(), # initial samples
                y0=-Y_init.ravel())
# fit GP model to samples for plotting
gpr.fit(r.x_iters, -r.func_vals)
# Plot the fitted model and the noisy samples
plot_approximation(gpr, X_plot, Y_plot, r.x_iters, -r.func_vals, show_
legend=True)
```

Running the preceding code will generate Figure 3-13, which shows a concentration of samples around the global maximum at the left peak. Note that the samples are not exactly the same as in our previous example due to the nondeterministic nature of the optimization procedure as well as the randomness in the observation model.

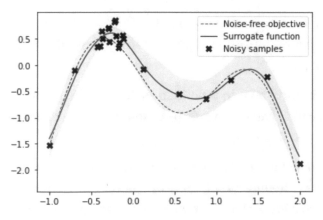

Figure 3-13. *Visualizing the proposed samples using the gp_minimize() function*

We can also show the plots on the distances of consecutive proposals and the best-observed value. As shown in Figure 3-14, even though the optimizer obtains a high value at the second iteration, it continues to explore promising regions with high uncertainty, as indicated by the two peaks in the distance plot.

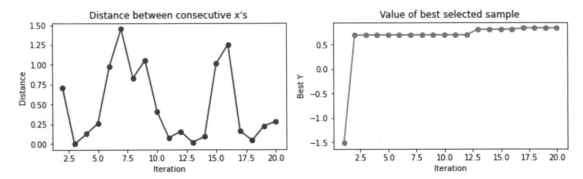

Figure 3-14. *Visualizing the convergence plots*

Summary

Bayesian optimization is an extension of the classic Bayesian decision theory. The extension goes into its use and choice of surrogate and acquisition functions. In this chapter, we covered the following list of items:

- Bayesian optimization requires defining a utility function that measures the value of the returned dataset in seeking the global optimum.

- The inner BO loop involves seeking the location that maximizes the acquisition function, and the outer BO loop seeks the location of the global optimum.

- The acquisition function is defined as the expected marginal gain in utility, which can be myopic (one-step lookahead) or nonmyopic (multi-step lookahead).

- Optimizing the multi-step lookahead expected marginal gain in utility follows Bellman's principle of optimality and can be expressed as a recursive form, that is, a sum of the immediate expected marginal gain in utility and the maximal expected marginal gain from all future evolutions.

- Expected improvement is a widely used one-step lookahead acquisition function that recommends the best-observed value when optimization terminates and has a friendly closed-form expression to support fast computation.

In the next chapter, we will revisit the Gaussian process and discuss GP regression using a widely used framework: GPyTorch.

## CHAPTER 4

Gaussian Process Regression with GPyTorch

So far, we have grasped the main components of a Bayesian optimization procedure: a surrogate model that provides posterior estimates on the mean and uncertainty of the underlying objective function and an acquisition function that guides the search for the next sampling location based on its expected gain in the marginal utility. Efficiently calculating the posterior distributions becomes essential in the case of parallel Bayesian optimization and Monte Carlo acquisition functions. This branch evaluates multiple points simultaneously discussed in a later chapter.

In this chapter, we will focus on the first component of building and refreshing a surrogate model and its implementation based on a highly efficient, state-of-the-art package called GPyTorch. The GPyTorch package is built on top of PyTorch and inherits all its built-in advantages, such as GPU acceleration and auto-differentiation. It serves as the backbone of BoTorch, the state-of-the-art BO package widely used in research and practice and to be introduced in the next chapter.

Introducing GPyTorch

Gaussian processes (GP) model the underlying objective function as a sample from the distribution of functions, where the distribution takes a prior form and gets updated as a posterior distribution upon revealing functional observations. Being a nonparametric model, it has an unlimited expressive capacity to learn from the training data and make predictions with a quantified uncertainty level. After choosing a kernel function to encode our prior assumption on the smoothness of the objective function, we can fit a GP model and apply it in a regression setting to obtain the posterior mean and variance at any new point, a topic covered in Chapter 2.

© Peng Liu 2023
P. Liu, *Bayesian Optimization*, https://doi.org/10.1007/978-1-4842-9063-7_4

When we are concerned with estimating the value at a new location along with its associated variance, performing either interpolation or extrapolation, we are conducting GP regression. Depending on the nature of the outcome, the model can be extended to the classification setting as well. In the next section, we will briefly cover the basics of PyTorch that underpins the GPyTorch library, followed by a short review of the basics of GP regression and an implementation case using GPyTorch.

The Basics of PyTorch

Backed by the Torch library, PyTorch has become one of the most popular frameworks used for deep learning tasks. It is widely used among researchers and practitioners in machine learning due to its ease of use, dynamic computational graph, and the "Pythonic" nature than other alternative frameworks such as TensorFlow. That is, if you are comfortable writing some Python code and checking the immediate output before moving on, then coding in PyTorch works much the same way. The code chunks in PyTorch are examinable and interruptible, making it convenient for debugging and packaging.

The essential component of PyTorch is its *tensor* data structure, which acts like a NumPy ndarray (multidimensional array) in a CPU but lives in a GPU instance during computation, which makes it much faster for specific tasks that benefit from graphical processing. It also supports *automatic differentiation*, which enables the automatic calculation of gradients when performing gradient descent optimization using backpropagation in neural networks. Such efficient and exact differentiation (assuming the objective function is differentiable) makes it much easier and faster to build and train a modern neural network model, which could be further boosted via parallel computation.

Let us look at a simple example to illustrate the auto-differentiation feature, or *autograd*. First, we create a tensor object from a list as follows:

```
>>> import torch
>>> a = torch.tensor([1,2,3])
>>> print(a)
tensor([1, 2, 3])
```

Note that the data type in the output may differ depending on the specification of the environment. The output suggests that the variable a is a tensor object. Alternatively, we can also create a tensor by converting a NumPy array:

```
>>> b = np.array([2,3,4])
>>> b = torch.from_numpy(b)
>>> print(b)
tensor([2, 3, 4])
```

Common NumPy operations such as addition and multiplication also have corresponding functions in PyTorch, often following the same naming convention. For example, we can add a and b using torch.add as follows:

```
>>> c = torch.add(a, b)
>>> print(c)
tensor([3, 5, 7])
```

When creating a tensor objective, we can specify requires_grad=True to enable the autograd feature. This means that in all follow-up computation that starts from this variable, the history of computation in all resulting outputs will be accumulated. Let us create another variable a_grad, where we use floating numbers since autograd does not work with integers:

```
>>> a_grad = torch.tensor([1.0,2.0,3.0], requires_grad=True)
>>> print(a_grad)
tensor([1., 2., 3.], requires_grad=True)
```

Next, we will create another variable by squaring a_grad:

```
>>> c = a_grad**2
>>> print(c)
tensor([1., 4., 9.], grad_fn=<PowBackward0>)
```

Here, the grad_fn attribute suggests that this is a power function when performing backward propagation and computing the gradients. Recall that in basic calculus, for a power function $y = x^2$, the gradient (or first derivative) is $y' = 2x$. Let us sum up all the elements to aggregate the inputs into a single output:

```
>>> out = c.sum()
>>> print(out)
tensor(14., grad_fn=<SumBackward0>)
```

Other than the summed result of 14, the output also suggests that this result comes from a summation operation. With all the basic operations specified and configured in the computation graph, we can perform backpropagation using the autograd feature by invoking the backward() function:

```
>>> out.backward()
>>> print(a_grad.grad)
tensor([2., 4., 6.])
```

The gradients are now automatically calculated and accessible via the grad attribute, which matches our expected output.

When performing the inner optimization in a BO loop, that is, locating the maximum of the acquisition function, the gradient-based method is often used. When such gradients can be easily calculated, the optimization can be managed efficiently. Methods such as multi-start stochastic gradient ascent (as we want to maximize the acquisition function) heavily rely on the autograd feature in PyTorch, GPyTorch, and BoTorch in particular. We will discuss this algorithm in a later chapter.

Next, let us revisit the basics of GP regression.

Revisiting GP Regression

GP regression relies on the closed-form solution for the predictive mean and variance of the random variable y_* at any new sampling location $\mathbf{x}_*$. Suppose we have collected a training set of n observations $\mathcal{D}_n = \{(\mathbf{x}_i, y_i)\}_{i=1}^{n}$, where each sampled location $\mathbf{x}_i \in \mathbb{R}^d$ is a d-dimensional feature vector in compact set $\mathcal{X} \subset \mathbb{R}^d$, and $y_i \in \mathbb{R}$ is a noise-perturbed scalar observation with $y_i = f(\mathbf{x}_i) + \epsilon_i$. We often assume a standard Gaussian noise with a homogenous variance across the whole domain, that is, $\epsilon_i \sim \mathcal{N}(0, \sigma^2)$.

Under the GP framework, the estimated objective is characterized by a mean function $\mu : \mathcal{X} \to \mathbb{R}$ and a covariance function $k : \mathcal{X} \times \mathcal{X} \to \mathbb{R}$, where $\mu(\mathbf{x}) = \mathbb{E}[f(\mathbf{x})]$ and $k(\mathbf{x}, \mathbf{x}') = \mathbb{E}\left[\left(f(\mathbf{x}) - \mathbb{E}[f(\mathbf{x})]\right)\left(f(\mathbf{x}') - \mathbb{E}[f(\mathbf{x}')]\right)\right]$. For any finite set of random variables within the domain, their joint distribution follows a multivariate Gaussian distribution, resulting in $y_{1:n} \sim \mathcal{N}(\mathbf{0}, \mathbf{K}_n)$, where we assume a constant zero mean function $\mu = 0$ and $\mathbf{K}_n = \mathbf{K}_n(\mathbf{x}_{1:n}, \mathbf{x}_{1:n})$ denotes the covariance matrix, noting $\mathbf{K}_n(\mathbf{x}_{1:n}, \mathbf{x}_{1:n})_{i,j} = k(\mathbf{x}_i, \mathbf{x}_j)$.

The predictive distribution for y_* at a new location $\mathbf{x}_*$ is again normally distributed, that is, $p(y_*; \mathbf{x}_*; \mathcal{D}_n) = \mathcal{N}(y_* | \mu_*, \sigma_*^2)$, where $\mu_* = \mathbf{k}(\mathbf{x}_*, \mathbf{x}_{1:n})^T(\mathbf{K}_n + \sigma^2 \mathbf{I})^{-1}y_{1:n}$ and $\sigma_*^2 = k(\mathbf{x}_*, \mathbf{x}_*) - \mathbf{k}(\mathbf{x}_*, \mathbf{x}_{1:n})^T(\mathbf{K}_n + \sigma^2 \mathbf{I})^{-1}\mathbf{k}(\mathbf{x}_*, \mathbf{x}_{1:n})$, and $\mathbf{k}(\mathbf{x}_*, \mathbf{x}_{1:n}) = [k(\mathbf{x}_*, \mathbf{x}_*), \ldots, k(\mathbf{x}_*, \mathbf{x}_*)]^T$.

Now, we will look at how to build a GP regression model using GPyTorch.

Building a GP Regression Model

Gaussian processes offer us a tool to make inferences on unobserved random variables. The inference is exact when the posterior probability distribution has a closed-form expression. However, when the exact inference is computationally intractable, such as computing high-dimensional integrals, the approximate inference is often used as a trade-off between accuracy and speed. When the posterior distribution is difficult to compute, sample, or both, we can use a biased but simpler approximation to carry out the inference.

Figure 4-1 illustrates these two approaches when making inferences or predictions for a new set of input points.

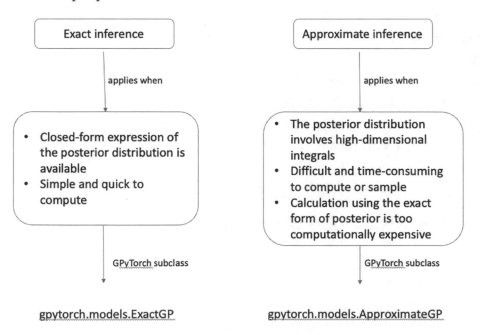

Figure 4-1. *Comparing exact and approximate inference in GPyTorch*

Since the objective function, say $f(x) = -\cos(\pi x) + \sin(4\pi x)$, is a relatively simple one-dimensional function, the exact GP inference by inheriting from gpytorch.models. ExactGP suffices in this working example.

In the following code listing, we first define the objective function, which is then used to generate ten noisy observations within the range of 0 and 1. Note that we have used torch functions such as cos(), sin(), randn(), and linspace(), all of which correspond to NumPy functions of the same name. We set the noise level to 0.1. When plotting the ten noisy observations, both variables X_init and y_init are converted to NumPy using the built-in numpy() function.

Listing 4-1. Generating noisy observations

```
# define the objective function
def f(x, noise=0):
    return -torch.cos(np.pi*x) + torch.sin(4*np.pi*x) + noise*torch.
    randn(*x.shape)
# observation noise
noise = 0.1
# number of observations
N = 10
# initial observations upon initiation
X_init = torch.linspace(0.05, 0.95, N)
y_init = f(X_init, noise=noise)
# plot noisy observations
plt.figure(figsize=(8, 6))
plt.plot(X_init.numpy(), y_init.numpy(), 'o');
```

Running the code generates Figure 4-2.

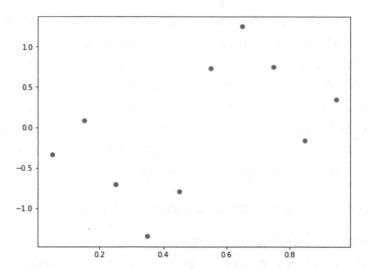

Figure 4-2. *Visualizing the initial ten noisy observations*

Before fitting a GP model, we must define the mean and covariance (kernel) functions for the prior GP. There are multiple mean functions available in GPyTorch, including the zero mean function gpytorch.means.ZeroMean(), the constant mean function gpytorch.means. ConstantMean(), and the linear mean function gpytorch.means.LinearMean(). In this case, we use the constant mean function, assuming that there is no systematic trend in the data.

As for the kernel function, we choose one of the most widely used kernels: the RBF kernel, or the squared exponential kernel $k(x_1,x_2)=\exp\left(-\dfrac{\|x_1-x_2\|^2}{2l^2}\right)$, where l is the length-scale parameter that can be optimized via a maximum likelihood procedure. The RBF kernel can be instantiated by calling the class gpytorch.kernels.RBFKernel(). We would also add a scaling coefficient to the output of the kernel function, simply by wrapping it with kernels.ScaleKernel(). Indeed, as the GPyTorch official documentation suggests: If you don't know what kernel to use, we recommend that you start out with a gpytorch.kernels.ScaleKernel(gpytorch.kernels.RBFKernel). The following code listing defines the mean and kernel functions.

Listing 4-2. Defining the mean and kernel functions

```
mean_fn = gpytorch.means.ConstantMean()
kernel_fn = gpytorch.kernels.ScaleKernel(gpytorch.kernels.RBFKernel())
```

Next, we will define a class to perform GP regression based on the exact GP inference class gpytorch.models.ExactGP. As shown in the following code listing, we define a class called GPRegressor, which inherits from gpytorch.models.ExactGP and is used to perform simple GP inference. There are three functions in the class: __init__(), forward(), and predict(). The __init__() function is triggered upon instantiating the class and stores the three input attributes: the prior mean and functions, as well as the likelihood function. In the case of GP regression, since we assume a noise-perturbed observation model, the Gaussian likelihood function maps the underlying functional distribution $p(f|\mathbf{x})$ to the marginal distribution $p(y|\mathbf{x})$, where $y = f(\mathbf{x}) + \epsilon$ is a noisy observation for an arbitrary location $\mathbf{x}$, and $\epsilon \sim \mathcal{N}\left(0, \sigma^2\right)$.

Listing 4-3. Defining a customized class for GP inference

```
class GPRegressor(gpytorch.models.ExactGP):
    def __init__(self, X, y, mean, kernel, likelihood=None):
        # choose the standard observation model as required by exact GP
        inference
        if likelihood is None:
            likelihood = gpytorch.likelihoods.GaussianLikelihood()
        # initiate the superclass ExactGP to refresh the posterior
        super().__init__(X, y, likelihood)
        # store attributes
        self.mean = mean
        self.kernel = kernel
        self.likelihood = likelihood

    # compute the posterior distribution after conditioning on the
    training data
    def forward(self, x):
        mean_x = self.mean(x)
        covar_x = self.kernel(x)
        # return a posterior multivariate normal distribution
        return gpytorch.distributions.MultivariateNormal(mean_x, covar_x)

    # compute the marginal predictive distribution of y given x
    def predict(self, x):
        # set to evaluation mode
        self.eval()
        # perform inference without gradient propagation
```

```
with torch.no_grad():
    # get posterior distribution p(f|x)
    pred = self(x)
    # convert posterior distribution p(f|x) to p(y|x)
    return self.likelihood(pred)
```

Next, the forward() function computes the posterior distribution at any input location and returns a multivariate normal distribution parameterized by the realization of the posterior mean and kernel functions at the particular location. This function gets called automatically when an instance of the GPRegressor class is used to make predictions. Note that MultivariateNormal is the only distribution allowed in the case of exact GP inference in GPyTorch.

Finally, the predict() function takes the output $p(f|\mathbf{x})$ from forward() and overlays an observation model to produce $p(y|\mathbf{x})$. Note that we need to set the GP model to the inference mode via the eval() function and performance inference without calculating the gradients, as indicated by the torch.no_grad() context manager.

We can then create a model by passing in the required arguments in the self-defined class as follows, which will train a GP model based on the provided training set and model configurations for the mean and kernel functions:

```
>>> model = GPRegressor(X_init, y_init, mean_fn, kernel_fn)
```

Now, let us visualize the posterior GP after fitting the model to the ten noisy observations. As shown in the following code listing, we create a function that accepts the learned GP model instance and optionally the range of the x-axis stored in xlim for plotting. We first extract the training data by accessing the train_inputs and train_targets attributes, respectively, both converted to NumPy format. When xlim is not provided, we rely on the upper and lower limits of X_train with a slight extension. Next, we create a list of equally spaced input locations in X_plot and score them to obtain the corresponding posterior predictive distributions, whose mean, upper, and lower bounds are used for plotting.

Listing 4-4. Plotting the fitted GP regression model

```
def plot_model(model, xlim=None):
    """

    Plot 1D GP model
    we
```

```
input:
model : gpytorch.models.GP
xlim : the limit of x axis, tuple(float, float) or None
"""
# extract training data in numpy format
X_train = model.train_inputs[0].cpu().numpy()
y_train = model.train_targets.cpu().numpy()
# obtain range of x axis
if xlim is None:
    xmin = float(X_train.min())
    xmax = float(X_train.max())
    x_range = xmax - xmin
    xlim = [xmin - 0.05 * x_range,
            xmax + 0.05 * x_range]
# create a list of equally spaced input locations following the same
dtype as parameters
model_tensor_example = list(model.parameters())[0]
X_plot = torch.linspace(xlim[0], xlim[1], 200).to(model_tensor_example)
# generate predictive posterior distribution
model.eval()
predictive_distribution = model.predict(X_plot)
# obtain mean, upper and lower bounds
lower, upper = predictive_distribution.confidence_region()
prediction = predictive_distribution.mean.cpu().numpy()
X_plot = X_plot.numpy()

plt.scatter(X_train, y_train, marker='x', c='k')
plt.plot(X_plot, prediction)
plt.fill_between(X_plot, lower, upper, alpha=0.1)
plt.xlabel('x', fontsize=14)
plt.ylabel('y', fontsize=14)
```

Applying this function to the model instance generates Figure 4-3.

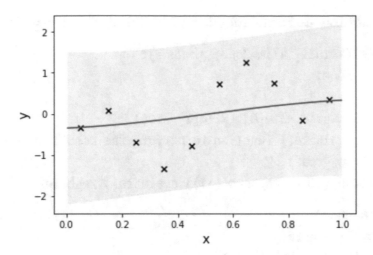

Figure 4-3. *Visualizing the fitted GP regression model*

We can see that the GP regression model is a rough fit with a wide confidence interval across the domain. Since the length-scale parameter for the kernel function has a direct influence on the shape of the resulting model fitting, we will discuss how to fine-tune this parameter to improve the predictive performance in the next section.

Fine-Tuning the Length Scale of the Kernel Function

Recall that we directly used the kernel function kernel_fn in the previous example without tuning its length-scale parameter. This parameter can be accessed via the lengthscale attribute, which can be manually set or optimized via the maximum likelihood procedure before entering the main GP loop.

In the following code listing, we change the value of the kernel's lengthscale parameter and observe the change in the shape, more specifically the smoothness, of the resulting kernel function. We create a total of six length-scale parameters stored in all_lengthscale, whose elements are sequentially iterated and passed into the lengthscale attribute of an RBF kernel function. We also create a plot_kernel() function to evaluate the kernel by comparing a dense list of input locations (stored in x) with the input location valued one (torch.ones((1))). Note that we need to trigger the kernel's evaluate() function to perform the evaluation of pairwise distances.

Listing 4-5. Plotting the kernel function with different length-scale parameters

```
def plot_kernel(kernel, xlim=None, ax=None):
    if xlim is None:
        xlim = [-3, 3]
    x = torch.linspace(xlim[0], xlim[1], 100)
    # call the evaluate() function to perform the actual evaluation
    with torch.no_grad():
        K = kernel(x, torch.ones((1))).evaluate().reshape(-1, 1)

    if ax is None:
        fig = plt.figure()
        ax = fig.add_subplot(111)
    ax.plot(x.numpy(), K.cpu().numpy())

k = gpytorch.kernels.RBFKernel()
all_lengthscale = np.asarray([0.2, 0.5, 1, 2, 4, 8])
figure, axes = plt.subplots(2, 3, figsize=(12, 8))
for tmp_lengthscale, ax in zip(all_lengthscale, axes.ravel()):
    k.lengthscale = tmp_lengthscale
    plot_kernel(k, ax=ax)
    ax.set_ylim([0, 1])
    ax.legend([tmp_lengthscale])
```

Running this code snippet generates Figure 4-4. When the length scale is small, the kernel function puts more weight on the nearby points and discards distance ones by setting the resulting similarity value to zero, thus displaying a less smooth curve. The kernel function becomes smoother as the length scale increases.

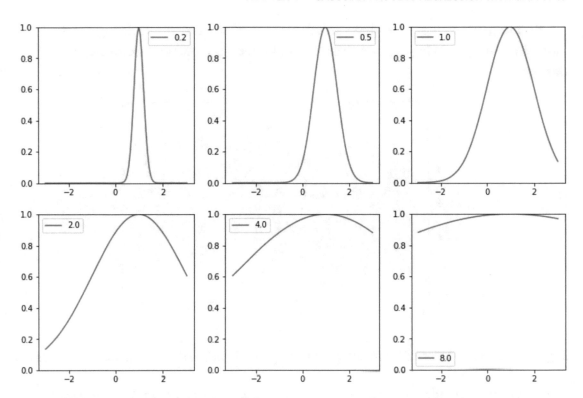

Figure 4-4. *Visualizing the kernel function under different length-scale parameter settings*

Back to our previous example. Upon initializing the kernel function in kernel_fn, we can access its length-scale parameter via the base_kernel.lengthscale attribute, where base_kernel refers to the RBF kernel wrapping in a scaling operation. As shown in the following, the default value is 0.6931, and the kernel object has a gradient attribute that will be used for gradient calculation that flows from the loss (negative maximum likelihood, to be discussed shortly) to the length-scale parameter:

```
>>> kernel_fn.base_kernel.lengthscale
tensor([[0.6931]], grad_fn=<SoftplusBackward0>)
```

Changing this parameter will result in a different GP model fitting. For example, we can manually set the length-scale parameter to 0.1 via the following code snippet:

```
>>> kernel_fn.base_kernel.lengthscale = 0.1
>>> model = GPRegressor(X_init, y_init, mean_fn, kernel_fn)
>>> plot_model(model)
```

Running this code snippet generates Figure 4-5, where the model fitting now looks better in terms of varying the overall confidence intervals along with the observed data points.

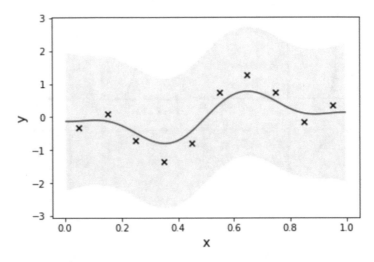

Figure 4-5. *GP regression model after setting the length-scale parameter manually*

However, manually setting the length-scale parameter and observing the fitting quality is time-consuming and inefficient. We need to resort to an automatic procedure that seeks the optimal parameter by optimizing a specific metric, such as the likelihood of the observed data.

Given a GP model $f \sim \mathcal{GP}(\mu, k)$ and a training set $(\mathbf{X}, \mathbf{y})$, the likelihood of observing the outcome $\mathbf{y}$ can be expressed as

$$\mathcal{L} = p_f(\mathbf{y}|\mathbf{X}) = \int p(\mathbf{y}|f(\mathbf{X})) p(f(\mathbf{X})|\mathbf{X}) df$$

which can be computed exactly in an exact GP inference setting, that is, GP regression with a Gaussian likelihood model. Negating the joint likelihood gives rise to the loss to be minimized in the optimization procedure. In GPyTorch, this joint marginal likelihood can be computed via first creating a `gpytorch.mlls.ExactMarginalLogLikelihood()` instance, which takes the Gaussian likelihood and the exact GP model as inputs and outputs the exact marginal log likelihood (MLL) based on the posterior distribution $p(f|\mathbf{X})$ and the target observations $\mathbf{y}$.

Now let us look at how to optimize the length scale automatically based on an optimization procedure similar to maximum likelihood estimation (MLE), where the likelihood is negated and minimized. In the following code listing, we define a function to perform the training procedure based on the existing training dataset (X and y), the number of epochs (n_epochs, each epoch is a full pass of the training set), the learning rate (lr, used by the chosen optimizer), and optionally the training log indicator (verbose). Note that the variance σ^2 of the Gaussian noise term used in the Gaussian likelihood model is also another hyperparameter that needs to be fine-tuned. Both σ^2 and lengthscale can be accessed via the model.parameters() function.

Listing 4-6. Optimizing the hyperparameters

```
def train(model, X, y, n_epochs=100, lr=0.3, verbose=True):
    # switch to model training mode
    model.train()
    # Use the adam optimizer
    training_parameters = model.parameters()
    optimizer = torch.optim.Adamax(training_parameters, lr=lr)
    # initiate a GP model with Gaussian likelihood for MLL calculation
    mll = gpytorch.mlls.ExactMarginalLogLikelihood(model.likelihood, model)
    # start the training loop
    for e in range(n_epochs):
        # clear gradients to prevent gradient accumulation
        optimizer.zero_grad()
        # get posterior distribution p(f|x)
        out = model(X)
        # get negated exact marginal log likelihood p(y|x)
        loss = -mll(out, y)n
        # calculate gradients using autograd
        loss.backward()
        # perform gradient update
        optimizer.step()
        # print optimization log
        if verbose:
            if e % 5 == 0:
                lengthscale = model.kernel.base_kernel.lengthscale.
                squeeze(0).detach().cpu().numpy()
```

```
print(f"Epoch: {e}, loss: {loss.item():.3f}, lengthscale:
{lengthscale[0]:.3f}, noise: {model.likelihood.noise.
item():.3f}")
# print(f"Loss: {loss.item():.3f}, lengthscale:
{lengthscale}")
```

The function starts by switching the model to the training mode. As for the optimizer of the hyperparameters, we use the Adam optimizer, a popular choice known for its stable updates and quick convergence. We also create an exact GP likelihood instance that is used to calculate the joint marginal likelihood, an indication of the goodness of fit in each training iteration. Each step of the gradient descent update consists of clearing the existing gradients, calculating the loss function, backpropagating the gradients, and performing the weight update.

As shown in Figure 4-6, after training the model for a total of 100 epochs, the result GP fit looks much better, where the confidence bounds are much narrower compared with before. Examining the length scale after training completes gives a value of 0.1279, as shown in the following code snippet.

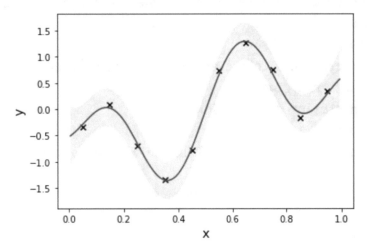

Figure 4-6. *Visualizing the GP fit after optimizing model hyperparameters, that is, the length scale and noise variance*

```
>>> kernel_fn.base_kernel.lengthscale
tensor([[0.1279]], grad_fn=<SoftplusBackward0>)
```

Next, we will look at the noise variance term as another hyperparameter that influences the smoothness of the fitted GP regression model.

Fine-Tuning the Noise Variance

The variance of the random noise can be accessed via the `likelihood.noise` attribute. For example, the optimal variance after optimization from the previous example is 0.0183:

```
>>> model.likelihood.noise
tensor([0.0183], grad_fn=<AddBackward0>)
```

We can also override the noise variance, which directly impacts the resulting smoothness and confidence interval of the fitted GP regression model. For example, Figure 4-7 shows the fitted GP model after setting the noise variance to 0.1, where the confidence intervals are much larger than earlier.

```
>>> model.likelihood.noise = 0.1
>>> plot_model(model)
```

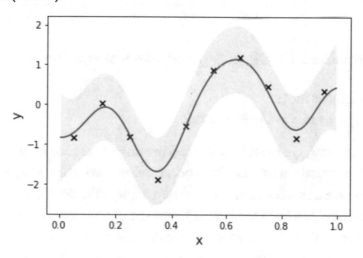

Figure 4-7. *Visualizing the fitted GP model after overriding the noise variance*

The automatic hyperparameter tuning procedure also performs well when the noise level increases in the previous example. In the following code listing, we increase the noise variance to 0.5 when generating noisy observations compared to 0.1 earlier. The rest of the code remains the same.

Listing 4-7. Generating observations with increased noise variance

```
y_init = f(X_init, noise=0.5)
model = GPRegressor(X_init, y_init, mean_fn, kernel_fn)
train(model, X_init, y_init)
plot_model(model)
```

Running the preceding code generates Figure 4-8. The confidence intervals are narrower and more adaptive than before. The optimized noise is now 0.0114, as shown in the following code snippet.

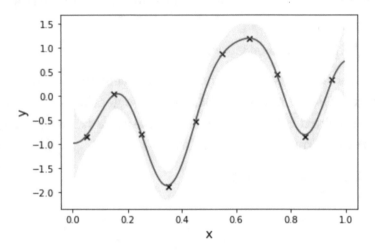

Figure 4-8. *Visualizing the GP model fitted with increased noise variance*

```
>>> model.likelihood.noise
tensor([0.0114], grad_fn=<AddBackward0>)
```

In certain situations, we would like to keep the noise variance fixed and unchanged throughout the optimization process. This occurs when we have some prior knowledge of the noise variance. In such a case, we can keep it out of the trainable parameters in the `training_parameters` from the `train()` earlier and only optimize the length scale. To achieve this, we can add an additional input argument `fixed_noise_variance` to accept the expected noise variance, if any. Next, replace the assignment of `training_parameters` using the following code snippet:

```
if fixed_noise_variance is not None:
    model.likelihood.noise = fixed_noise_variance
    training_parameters = [p for name, p in model.named_parameters() if
    not name.startswith('likelihood')]
else:
    training_parameters = model.parameters()
```

Here, we used the `named_parameters()` function to extract the parameters together with their names, which is used to filter out those parameters that start with the keyword "likelihood." We then train the model again using the new training procedure and obtain the result in Figure 4-9 with a fixed noise variance of 0.1.

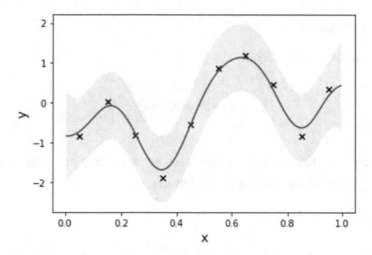

Figure 4-9. *Fitting GP regression model with fixed noise variance and optimized length scale*

Feel free to check out the entire training code in the accompanying notebook at `https://github.com/jackliu333/bayesian_optimization_theory_and_practice_with_python/blob/main/Chapter_4.ipynb`.

The previous example shows the importance of choosing a proper kernel function and optimizing the hyperparameters to obtain a good GP fit. In the next section, we will look at different kernel functions available in GPyTorch and their combined usage in improving model fitting performance.

Delving into Kernel Functions

There are multiple kernel functions available in GPyTorch. Popular choices include the RBF (radial basis function) kernel, rational quadratic kernel, Matern52 kernel, linear kernel, polynomial kernel, and periodic kernel. The definition of each kernel function is given in the following, where we have multiplied the noise variance σ^2 as the scaling factor to determine the average distance of the function away from its mean:

- RBF kernel

$$k(x_1,x_2)=\sigma^2\exp\left(-\frac{r^2}{2l^2}\right)$$

where $r=\|x_1-x_2\|$.

- Rational quadratic kernel

$$k(x_1,x_2)=\sigma^2\exp\left(1+\frac{r^2}{2al^2}\right)^{-a}$$

where the additional hyperparameter a determines the relative weighting of large-scale and small-scale variations.

- Matern52 kernel

$$k(x_1,x_2)=\sigma^2\left(1+\sqrt{5}\frac{r}{l}+\frac{5}{3}\frac{r^2}{l^2}\right)\exp\left(-\sqrt{5}\frac{r}{l}\right)$$

- Linear kernel

$$k(x_1,x_2)=\sum_i\sigma_i^2 x_{1,i}x_{2,i}$$

- Polynomial kernel

$$k(x_1,x_2)=\sigma^2\left(x_1x_2+c\right)^d$$

- Periodic kernel

$$k(x_1,x_2)=\sigma^2\exp\left(-2\frac{\sin^2(\pi r)}{l^2}\right)$$

In the following code listing, we plot each kernel function using the `plot_kernel()` function defined earlier. Note that we used the string match function `split()` to extract the kernel function name as the title for each subplot.

Listing 4-8. Visualizing different kernel functions

```
covariance_functions = [gpytorch.kernels.RBFKernel(),
                        gpytorch.kernels.RQKernel(),
                        gpytorch.kernels.MaternKernel(nu=5/2),
                        gpytorch.kernels.LinearKernel(power=1),
                        gpytorch.kernels.PolynomialKernel(power=2),
                        gpytorch.kernels.PeriodicKernel()
                        ]
figure, axes = plt.subplots(2, 3, figsize=(9, 6))
axes = axes.ravel()
for i, k in enumerate(covariance_functions):
    plot_kernel(k, ax=axes[i])
    axes[i].set_title(str(k).split('(')[0])
figure.tight_layout()
```

Running the preceding code will generate Figure 4-10. Although each type of kernel displays different characteristics, we can further combine them via addition and multiplication operations to create an even more flexible representation, as we will show in the next section.

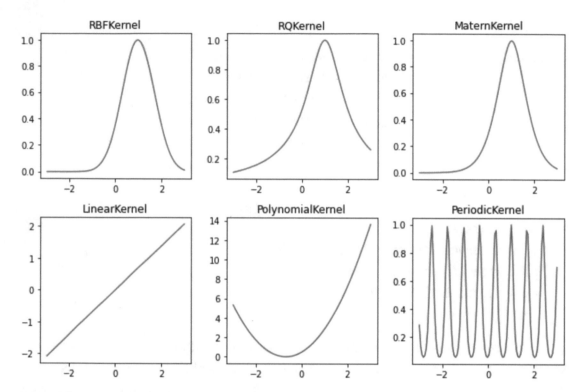

Figure 4-10. *Visualizing different kernel functions*

Combining Kernel Functions

Kernel functions can be combined using the addition operator + or the multiplication operator * in GPyTorch. Both operations are illustrated as follows:

- Deriving a composite kernel function by adding two kernel functions

$$k(x_1,x_2) = k_1(x_1,x_2) + k_2(x_1,x_2)$$

- Deriving a composite kernel function by multiplying two kernel functions

$$k(x_1,x_2) = k_1(x_1,x_2)k_2(x_1,x_2)$$

Now let us look at how to combine different kernel functions. We first create a list `kernel_functions` to store three kernel functions: linear kernel, periodic kernel, and RBF kernel. We also overload the + and * operators as lambda functions, so that +(a,b) becomes a+b and *(a,b) is equivalent to a*b. Since we would like to apply the two operators for each unique combination of any two kernel functions, the `combinations()` function from the `itertools` package is used to generate the unique combinations to be iterated over. See the following code listing for the full implementation details.

Listing 4-9. Combining different kernel functions

```
kernel_functions = [gpytorch.kernels.LinearKernel(power=1),
                    gpytorch.kernels.PeriodicKernel(),
                    gpytorch.kernels.RBFKernel()]
# overload the + and * operators
operations = {'+': lambda x, y: x + y,
              '*': lambda x, y: x * y}
figure, axes = plt.subplots(len(operations), len(kernel_functions),
figsize=(9, 6))

import itertools
axes = axes.ravel()
count = 0
# iterate through each unique combinations of kernels and operators
for j, base_kernels in enumerate(itertools.combinations(kernel_
functions, 2)):
    for k, (op_name, op) in enumerate(operations.items()):
        kernel = op(base_kernels[0], base_kernels[1])
        plot_kernel(kernel, ax=axes[count])
        kernel_names = [
            str(base_kernels[i]).split('(')[0] for i in [0, 1]
        ]
        axes[count].set_title('{} {} {}'.format(kernel_names[0], op_name,
                                                kernel_names[1]),
                                                fontsize=12)
        count += 1
figure.tight_layout()
```

Running the preceding code will generate Figure 4-11. Note that when adding the linear kernel with the periodic kernel, the resulting combined kernel displays both linear trend and periodicity.

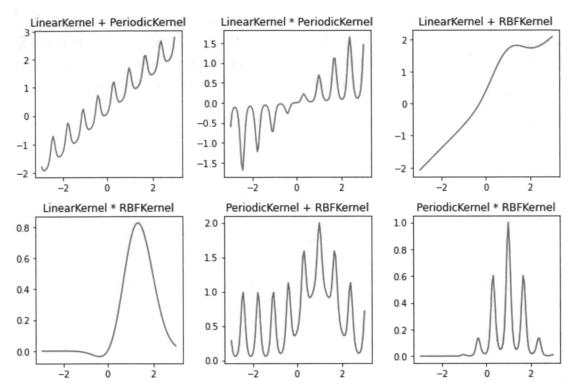

Figure 4-11. *Visualizing different combined kernels*

Of course, we can extend the combination to more than two kernels or even apply an automatic procedure that searches for the optimal combination of kernels based on the available training data. The next section will demonstrate the model's learning capacity improvement by applying more complex and powerful kernel combinations.

Predicting Airline Passenger Counts

The airline passenger count data records the number of passengers on monthly international flights from 1949 to 1960. The following code listing loads the dataset and splits it into training and test sets, with the middle 20 years of data allocated to the test set.

Listing 4-10. Loading the airline count dataset

```
data = np.load('airline.npz')
X = torch.tensor(data['X'])
y = torch.tensor(data['y']).squeeze()
train_indices = list(range(70)) + list(range(90, 129))
```

```
test_indices = range(70, 90)
X_train = X[train_indices]
y_train = y[train_indices]
X_test = X[test_indices]
y_test = y[test_indices]
plt.figure(figsize=(5, 3))
plt.plot(X_train.numpy(), y_train.numpy(), '.')
```

Running the preceding code generates Figure 4-12, showing the training set to be used to fit a GP regression model. The figure suggests an increasing trend with a seasonal pattern, which can be explained by the growing travel demand and seasonality across the year. In addition, the vertical range is also increasing, indicating a heteroscedastic variance across different months. These structural components, namely, the trend, seasonality, and random noise, are essential elements often modeled in the time-series forecasting literature. In this exercise, we will explore different kernel combinations and eventually model the GP using three types of composite kernels in an additive manner.

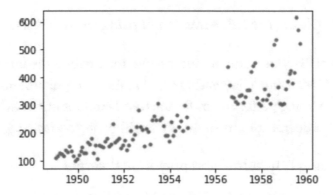

Figure 4-12. *Visualizing the training set of the monthly airline passenger count data*

Let us try a vanilla RBF kernel function and observe the fitting performance. The following code listing uses a similar hyperparameter optimization procedure. The only notable difference is the use of the double() function to convert the GP model instance to double precision format.

Listing 4-11. GP fitting using RBF kernel alone

```
mean_fn = gpytorch.means.ConstantMean()
kernel_fn = gpytorch.kernels.ScaleKernel(gpytorch.kernels.RBFKernel())
model = GPRegressor(X_train, y_train, mean_fn, kernel_fn).double()
```

```
train(model, X_train, y_train)
plot_model(model, xlim = [1948, 1964])
```

Running this code listing outputs Figure 4-13, which shows a poor fitting performance in that the structural components are not captured.

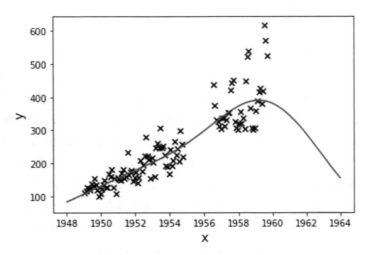

Figure 4-13. *Visualizing the GP regression fit using the RBF kernel*

The fitted model displays a smooth increasing trend within the interpolation region (between 1949 and 1960) and a decreasing trend in the extrapolation region (after 1960). We can improve the fitting performance for the trend component by adding another polynomial kernel of degree one, as shown in the following code listing.

Listing 4-12. Adding RBF kernel and polynomial kernel

```
kernel_fn = gpytorch.kernels.ScaleKernel(gpytorch.kernels.
PolynomialKernel(power=1)) + \
gpytorch.kernels.ScaleKernel(gpytorch.kernels.RBFKernel())
model = GPRegressor(X_train, y_train, mean_fn, kernel_fn).double()
train(model, X_train, y_train, verbose=False)
plot_model(model, xlim = [1948, 1964])
```

Running the preceding code generates Figure 4-14, where the long-term trend displays a certain level of flexibility compared with earlier.

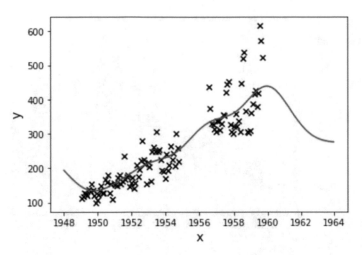

Figure 4-14. *Visualizing the GP regression fit using the RBF kernel and polynomial kernel*

Since the data is seasonal, we can additionally build another kernel to capture such seasonality. In the following code listing, we create a seasonality-seeking kernel called k_seasonal by adding the periodic kernel, linear kernel, and RBF kernel. This kernel will then work together with the previous trend-seeking kernel via the AdditiveKernel() function, a utility function that supports summing over multiple component kernels.

Listing 4-13. Modeling trend and seasonality

```
k_trend = (gpytorch.kernels.ScaleKernel(gpytorch.kernels.
PolynomialKernel(power=1)) +
          gpytorch.kernels.ScaleKernel(gpytorch.kernels.RBFKernel()))
k_seasonal = (gpytorch.kernels.ScaleKernel(gpytorch.kernels.
PeriodicKernel()) *
             gpytorch.kernels.LinearKernel() *
             gpytorch.kernels.ScaleKernel(gpytorch.kernels.RBFKernel()))
kernel_fn = gpytorch.kernels.AdditiveKernel(k_trend, k_seasonal)
gpytorch.kernels.ScaleKernel(gpytorch.kernels.RBFKernel())
model = GPRegressor(X_train, y_train, mean_fn, kernel_fn).double()
train(model, X_train, y_train, verbose=False)
plot_model(model, xlim = [1948, 1964])
```

Running the preceding code generates Figure 4-15, where the trend and seasonality components are well captured.

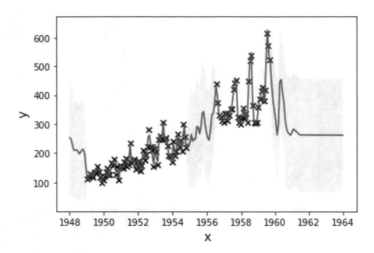

Figure 4-15. *Visualizing the GP regression fit using composite kernels on both trend and seasonality*

However, the fitting still appears wiggly due to a uniform noise variance applied across the whole domain. In the next attempt, we will add another kernel that increases the noise variance as we travel from left to right. Such kernel can be considered as a multiplication of a white noise kernel and a linear kernel, where the white noise kernel assumes a constant variance across the whole domain, and the linear kernel gradually lifts the variance when multiplied by the white noise kernel.

In the following code listing, we first define a class as the white noise kernel. When calling the class instance, its forward() function will return the prespecified noise variance (wrapped via a lazy tensor format for efficiency) when the two input locations overlap and zero otherwise. This kernel is then multiplied with a linear kernel to model the heteroscedastic noise component, which is added to the previous two composite kernels to model the GP jointly.

Listing 4-14. Modeling trend, seasonality, and noise

```python
from gpytorch.lazy import DiagLazyTensor
class WhiteNoiseKernel(gpytorch.kernels.Kernel):
    def __init__(self, noise=1):
        super().__init__()
        self.noise = noise

    def forward(self, x1, x2, **params):
        if torch.equal(x1, x2):
```

```
        return DiagLazyTensor(torch.ones(x1.shape[0]).to(x1) *
        self.noise)
    else:
        return torch.zeros(x1.shape[0], x2.shape[0]).to(x1)

k_noise = gpytorch.kernels.ScaleKernel(WhiteNoiseKernel(noise=0.1)) *
gpytorch.kernels.LinearKernel()
kernel_fn = gpytorch.kernels.AdditiveKernel(k_trend, k_seasonal, k_noise)
gpytorch.kernels.ScaleKernel(gpytorch.kernels.RBFKernel())
model = GPRegressor(X_train, y_train, mean_fn, kernel_fn).double()
train(model, X_train, y_train, verbose=False)
plot_model(model, xlim = [1948, 1964])
```

Running the preceding code generates Figure 4-16, where the GP fitting looks much better now. The overall trend, seasonality, and heteroscedastic noise are well captured in the interpolation region and continued in the extrapolation region.

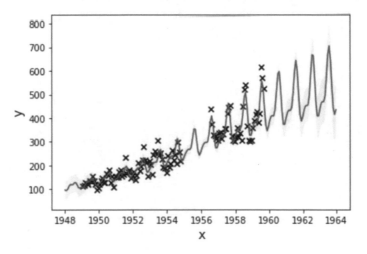

Figure 4-16. *Visualizing the GP regression fit using composite kernels on the trend, seasonality, and heteroscedastic noise*

Summary

Building an excellent surrogate model is essential for effective BO. In this chapter, we discussed one of the most popular (and modular) GP packages, GPyTorch, and its use in various aspects. Specifically, we covered the following:

- The fundamentals of coding in PyTorch and its autograd feature

- Step-by-step implementation of exact GP inference

- Optimizing GP hyperparameters for the kernel function and noise variance by minimizing the negated MLE

- Forming composite kernels for more effective GP model fitting

- A case study on developing the approximate kernel functions based on the characteristics of the dataset

In the next chapter, we will look at Monte Carlo acquisition functions and introduce a series of optimization techniques that turn out to be quite helpful in accelerating the BO process.

## CHAPTER 5

Monte Carlo Acquisition Function with Sobol Sequences and Random Restart

In the previous chapter, we introduced GPyTorch, the backbone horsepower used by BoTorch to obtain the GP posterior using PyTorch. This paves the way for our introduction to BoTorch, the main topic in this chapter. Specifically, we will focus on how it implements the expected improvement acquisition function covered in Chapter 3 and performs the inner optimization in search of the next best proposal for sampling location.

In addition, we will also cover another type of acquisition function that does not bear an analytic solution and can only be computed via the Monte Carlo approach. Examples include parallel Bayesian optimization, where more than one sampling location is proposed in each outer optimization, and nonmyopic acquisition functions, such as the multi-step lookahead variants. Understanding the estimation process of the functional evaluation of the acquisition function at a specific location using the Monte Carlo approach, including the quasi-Monte Carlo method, underpins our investigation of the core working mechanism of BoTorch from this chapter onward.

Analytic Expected Improvement Using BoTorch

There are multiple types of acquisition functions available in BoTorch. While a few acquisition functions, such as the expected improvement (EI), have an analytic form, the majority do not enjoy this property. We will examine the analytic EI in BoTorch in this

131

© Peng Liu 2023
P. Liu, *Bayesian Optimization*, https://doi.org/10.1007/978-1-4842-9063-7_5

section and switch to the corresponding Monte Carlo (MC)–based EI in the next section, which requires MC estimation for a functional evaluation of the acquisition function.

In the following section, we will introduce the Hartmann function, a widely used synthetic function to benchmark the performance of optimization algorithms.

Introducing Hartmann Function

The Hartmann function is a six-dimensional function with six local minima and one global minimum. Two things deserve our immediate attention. First, the domain of the function has six dimensions, which is more complex than the one-dimensional case we encountered in Chapter 3 and presents a more challenging state space for the search problem. Second, the functional shape is multimodal with multiple local minima, which requires the search algorithm to explore efficiently and avoid being stuck in such local minima.

The Hartmann function takes the following form:

$$f(\mathbf{x}) = -\sum_{i=1}^{4} \alpha_i \exp\left(-\sum_{j=1}^{6} A_{ij} \left(x_j - P_{ij} \right)^2 \right)$$

where α_i, A_{ij}, and P_{ij} are all constants. The function is usually evaluated over a six-dimensional unit hypercube, namely, $x_j \in (0, 1)$ for all $j = 1, ..., 6$. The global minimum is obtained at $\mathbf{x}^* = (0.20169, 0.150011, 0.476874, 0.275332, 0.311652, 0.6573)$, with $f(\mathbf{x}^*) = -3.32237$.

Let us look at how to play around with the Hartmann function available in BoTorch. First, we will need to install the package via the following command:

```
>>> !pip install botorch
```

Next, we will import it based on the test_function module from BoTorch. In addition, we fix the random seed for reproducibility purposes:

```
from botorch.test_functions import Hartmann
import numpy as np
import torch
import random

SEED = 8
random.seed(SEED)
```

```
np.random.seed(SEED)
torch.manual_seed(SEED)
```

We can create an instance of negated six-dimensional Hartmann function in a maximization setting. Inspecting the optimal value returns 3.32237.

Listing 5-1. Generating the Hartmann instance

```
>>> neg_hartmann6 = Hartmann(dim=6, negate=True)
>>> neg_hartmann6.optimal_value
3.32237
```

This instance will be used to provide noise-free functional evaluations. In the following, we will create ten random sampling locations following a uniform distribution in train_x and obtain the corresponding observations in train_obj, where we use the unsqueeze() function to convert it to a column vector:

```
# create 10x6 random numbers from a uniform distribution
train_x = torch.rand(10, 6)
# obtain the corresponding functional evaluation as a column vector
train_obj = neg_hartmann6(train_x).unsqueeze(-1)
```

Printing out train_obj gives the following result contained in a tensor:

```
>>> train_obj
tensor([[0.1066],
        [0.0032],
        [0.0027],
        [0.7279],
        [0.0881],
        [0.8750],
        [0.0038],
        [0.1098],
        [0.0103],
        [0.7158]])
```

Next, we will build a GP-powered surrogate model and optimize the associated hyperparameters using utilities provided by GPyTorch, a topic covered in the previous chapter.

GP Surrogate with Optimized Hyperparameters

When the underlying surrogate model is a Gaussian process, and there is only one objective value to be optimized, BoTorch provides the SingleTaskGP class that performs exact GP inference for the single-task setting. By default, it uses a GP prior with a Matérn kernel and a constant mean function. In the following, we instantiate a GP model with the initial training data generated earlier:

```
from botorch.models import SingleTaskGP
model = SingleTaskGP(train_X=train_x, train_Y=train_obj)
```

The model object now serves as the surrogate model and provides posterior mean and variance estimates for any sampling location within the unit hypercube.

The hyperparameters can be accessed via the named_hyperparameters() method as follows, where the output is wrapped in a list in order to print out the generator object:

```
>>> list(model.named_hyperparameters())
[('likelihood.noise_covar.raw_noise', Parameter containing:
  tensor([2.0000], requires_grad=True)),
 ('mean_module.constant', Parameter containing:
  tensor([0.], requires_grad=True)),
 ('covar_module.raw_outputscale', Parameter containing:
  tensor(0., requires_grad=True)),
 ('covar_module.base_kernel.raw_lengthscale', Parameter containing:
  tensor([[0., 0., 0., 0., 0., 0.]], requires_grad=True))]
```

Based on the output, the default value for the length scale is zero for all six dimensions and two for the noise variance. In order to optimize these hyperparameters using the maximum likelihood principle, we will first obtain an instance that contains the exact marginal log likelihood (stored in mll) and then use the fit_gpytorch_model function to optimize these hyperparameters of the GPyTorch model instance. The core optimization uses the L-BFGS-B procedure via the scipy.optimize.minimize() function. The following code snippet performs the optimization of the hyperparameters:

Listing 5-2. Optimizing the hyperparameters of the GP model

```
from gpytorch.mlls import ExactMarginalLogLikelihood
mll = ExactMarginalLogLikelihood(model.likelihood, model)
from botorch.fit import fit_gpytorch_model
fit_gpytorch_model(mll);
```

We can verify the hyperparameters now and check if these values have changed. As shown in the following, the hyperparameters now assume different values, all of which are optimized by maximizing the marginal log likelihood of the observed training data:

```
>>> list(model.named_hyperparameters())
[('likelihood.noise_covar.raw_noise', Parameter containing:
  tensor([0.0060], requires_grad=True)),
 ('mean_module.constant', Parameter containing:
  tensor([0.2433], requires_grad=True)),
 ('covar_module.raw_outputscale', Parameter containing:
  tensor(-2.1142, requires_grad=True)),
 ('covar_module.base_kernel.raw_lengthscale', Parameter containing:
  tensor([[-0.7155, -0.7190, -0.7218, -0.8089, -1.1630, -0.5477]],
         requires_grad=True))]
```

In the next section, we will use the analytic form of the expected improvement acquisition function and obtain the solution for the outer BO loop.

Introducing the Analytic EI

BoTorch provides the ExpectedImprovement class that computes the expected improvement over the current best-observed value based on the analytic formula we derived earlier. It requires two input parameters: a GP model which has a single outcome and optionally with optimized hyperparameters and the best-observed scalar value, assumed to be noiseless. The calculation is based on the posterior mean and variance of the GP surrogate model at a single specific sampling location.

Let us create an instance of ExpectedImprovement to calculate the EI at any given location. First, we will locate the best-observed value using the max() function and store the result in best_value. As shown in the following code snippet, the highest observed value is 0.875:

```
>>> best_value = train_obj.max()
>>> best_value
tensor(0.8750)
```

Next, we will instantiate the EI object based on best_value and the posterior GP model object. In the following code snippet, we pass the first observation to the EI object and obtain a result of 2.8202e-24, which is almost zero. Intuitively, this makes sense as the marginal utility from an already sampled location should be minimal.

```
>>> from botorch.acquisition import ExpectedImprovement
>>> EI = ExpectedImprovement(model=model, best_f=best_value)
>>> EI(train_x[0].view(1,-1))
tensor([2.8202e-24], grad_fn=<MulBackward0>)
```

We can further examine the implementation of the ExpectedImprovement class in BoTorch. In the following, we show the definition of the forward() method for this class, as given by the official GitHub page of BoTorch version 0.6.5:

Listing 5-3. Evaluating the EI acquisition function

```
from torch import Tensor
from torch.distributions import Normal

def forward(self, X: Tensor) -> Tensor:
    0# convert the data type of best observed value best_f to that of
    candidate X
    self.best_f = self.best_f.to(X)
    # obtain the posterior instance at candidate location(s) X
    posterior = self.model.posterior(
        X=X, posterior_transform=self.posterior_transform
    )
    # get posterior mean
    mean = posterior.mean
    # deal with batch evaluation and broadcasting
```

```
view_shape = mean.shape[:-2] if mean.shape[-2] == 1 else mean.
shape[:-1]
mean = mean.view(view_shape)
# get posterior standard deviation, floored to avoid dividing by zero
sigma = posterior.variance.clamp_min(1e-9).sqrt().view(view_shape)
# calculate the standardized unit improvement, with best_f broadcasted
if needed
u = (mean - self.best_f.expand_as(mean)) / sigma
# switch sign in the case of minimization
if not self.maximize:
    u = -u
# calculate ei based on analytic form
normal = Normal(torch.zeros_like(u), torch.ones_like(u))
ucdf = normal.cdf(u)
updf = torch.exp(normal.log_prob(u))
ei = sigma * (updf + u * ucdf)
return ei
```

In this function, we first use the to() method of a tensor object to align the data type of best_f to that of the candidate location X. We then access the posterior mean and variance of the candidate location, with the shape arranged according to batch evaluation and broadcasting needs and the standard deviation floored to avoid dividing by zero. With the unit improvement calculated in u, we can plug in the analytic form of EI, given as follows for ease of reference, and perform the calculation accordingly:

$$\alpha_{\mathrm{EI}} = \left(\mu - f^*\right)\Phi\left(\frac{\mu - f^*}{\sigma}\right) + \sigma\phi\left(\frac{\mu - f^*}{\sigma}\right)$$

where f^* is the best-observed value so far, and ϕ and Φ denote the probability and cumulative density function of a standard normal distribution, respectively. Note that the code extracts out a common factor of σ, making it easier for implementation.

Next, we will look at how to optimize the black-box Hartmann function using EI.

Optimization Using Analytic EI

Since most acquisition functions are nonconvex and often flat, such as EI, optimizing over these functions presents a nontrivial challenge. In such cases, it is important to devise an optimizer that is capable of jumping out of suboptimal candidate locations such as local optima or saddle points when searching for the global optimum.

The optimization horsepower is mainly provided by the `optimize_acqf()` function in BoTorch, which generates a set of candidate optimal locations using multi-start optimization based on stochastic gradient descent. In particular, the multi-start stochastic gradient descent methods are widely used for gradient-based stochastic global optimization in many machine learning and deep learning models. Let us briefly discuss this type of optimization method.

Stochastic gradient descent (SGD) is an optimization algorithm that performs gradient-based updates to the variable based on one or a mini-batch of training examples. However, SGD-based methods would struggle to locate the global optimum when the objective function is nonconvex and, instead, tend to converge to a saddle point or local optimum. Even when the objective function is convex, SGD-based methods would also have a challenge converging to the global optimum when the starting location is a disadvantageous one.

A common remedy to these challenges is performing SGD from multiple starting locations, selected randomly or according to some adaptive procedure. When the number of starting locations is sufficiently diverse, we could run multiple SGD optimizations in parallel and generate multiple search paths. Upon convergence, we will then report the most promising location, which could be identified as having the highest acquisition function value, as the result of optimization. In this way, it is more likely that our SGD-based methods will jump out of the saddle points or local optimum and find a global optimum. Figure 5-1 illustrates the multi-start stochastic gradient descent algorithm.

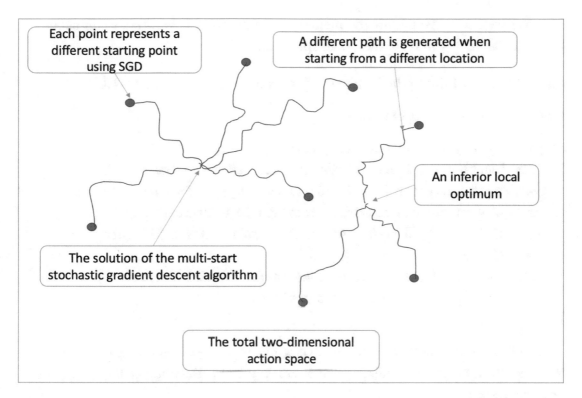

Figure 5-1. *Illustrating the multi-start stochastic gradient descent algorithm in a two-dimensional action space, where multiple starting points are selected to generate different paths in parallel. The final solution is obtained by aggregating all endpoints upon convergence and taking the best candidate. Note that additional projection to the nearest sampling location is needed in the case of discrete action space*

Using multi-start SGD, each restart can be treated as a separate optimization routine within a local neighborhood that aims at locating a locally optimal point. When aggregating multiple restarts and taking the best solution, we have a better chance of reaching the global optimum.

When using the optimize_acqf() function to perform multi-start optimization, the number of starting points is specified via the num_restarts argument. Besides, we also need to pass the bounds of the search space, a six-dimensional unit hypercube, via the bounds argument, and the number of raw examples for initialization via the raw_samples argument. Since we are only proposing one sampling location at each iteration, we also need to set q=1. Proposing more than one location in each iteration involves parallel Bayesian optimization, a topic to be covered in a later chapter.

After setting all these input arguments, we can run the following code snippet to generate the final solution to the global optimum:

Listing 5-4. Obtaining the best sampling location that maximizes EI

```
from botorch.optim import optimize_acqf

new_point_analytic, _ = optimize_acqf(
    acq_function=EI, # acquisition function to guide the search
    bounds=torch.tensor([[0.0] * 6, [1.0] * 6]), # 6d unit hypercube
    q=1, # generate one candidate location in each iteration
    num_restarts=20, # number of starting points for multistart
    optimization
    raw_samples=100, # number of samples for initialization
    options={}, # additional options if any
)
```

Printing out the candidate solution in new_point_analytic gives the following. Note that the associated acquisition values are ignored via the underscore sign in the returned result:

```
>>> new_point_analytic
tensor([[0.1715, 0.3180, 0.0816, 0.4204, 0.1985, 0.7727]])
```

BoTorch is characterized by its computational efficiency in optimizing different acquisition functions. In the following section, we will peek into the inner core of the optimize_acqf() function and learn more about its auto-differentiation-based optimization routine.

Grokking the Inner Optimization Routine

The default optimizer used by BoTorch is based on the scipy.optimize.minimize() function from the SciPy package and makes use of the L-BFGS-B routine. This is contained in the gen_candidates_scipy() function, which optimizes a prespecified acquisition function starting from a set of initial candidates. But first, let us look at how to set the starting conditions for the multi-start optimization routine. Since the underlying implementations are quite well wrapped, with multiple layers of functions calling each other, we will only highlight the main ones relevant to our context: optimizing an analytic EI and proposing a candidate search location sequentially.

Any optimization routine starts with an initial condition, the starting point of an optimizer, which could be set either randomly or following some heuristics. The initial conditions are set by the optional `batch_initial_conditions` argument in `optimize_acqf()`. Since this argument defaults to `None` and assuming we have no prior preference for generating the initial optimization condition, `optimize_acqf()` will use the `gen_batch_initial_conditions()` function to select a set of starting points in order to run the multi-start gradient-based algorithm.

These set of starting points, also referred to as initial conditions in `BoTorch`, are selected based on a multinomial probability distribution whose probabilities are determined by the value of the acquisition functions, additionally weighted by a temperature hyperparameter. Selecting the initial conditions concerns two steps: determining a much bigger set of random locations (specified by the `raw_samples` parameter) to cover the total action space as much as possible and selecting a subset of promising candidate locations to serve as initial starts (specified by the `num_restarts`). Let us look at these two steps in detail.

Intuitively, a good initial condition should bear a relatively high value when evaluated using the current acquisition function. This requires building a set of space-filling points so as to cover all regions of the search space and not leave out any big chunk of search space. Although these initial points are randomly selected, directly following a uniform distribution does not guarantee an even coverage of the whole search points using the set of the proposed random point. Instead, we would like a quasi-random sequence of points that has a low discrepancy in terms of distance among each other.

To this end, the Sobol sequences are used in `BoTorch` to form a uniform partition of the unit hypercube along each dimension. In other words, the quasi-Monte Carlo samples generated using Sobol sequences could fill the space more evenly, resulting in faster convergence and more stable estimates. The Sobol sequences are generated using the `torch.quasirandom.SobolEngine` class, which accepts three input arguments: `dimension` to specify the feature dimension, `scramble` to specify if the sequences would be scrambled, and `seed` to specify the seed for the random number generator.

We generate 200 random points of Sobol sequences in a two-dimensional unit hypercube in the following code snippet. First, we create a `SobolEngine` instance by specifying the corresponding input arguments and storing it in `sobol_engine`. We then call the `draw()` method to create 200 random points in `sobol_samples`, which are further decomposed into `xs_sobol` and `ys_sobol` to create the two-dimensional plot:

Listing 5-5. Generating Sobol sequences

```
from torch.quasirandom import SobolEngine
from matplotlib import pyplot as plt
plt.figure(figsize=(6,6))
sobol_engine = SobolEngine(dimension=2, scramble=True, seed=SEED)
sobol_samples = sobol_engine.draw(200)
xs_sobol = [x[0] for x in sobol_samples]
ys_sobol = [x[1] for x in sobol_samples]
plt.scatter(xs_sobol, ys_sobol)
```

Running this code snippet generates Figure 5-2. The points appear to be evenly distributed across the whole domain, providing a good space-filling design to "warm-start" the optimization procedure. A good initial design could also enable better estimates of, for example, the expectation operator and faster convergence to the global optimum using a specific optimization routine.

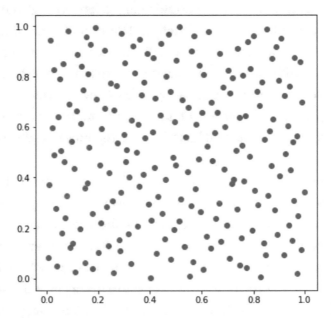

Figure 5-2. *Sobol sequences that consist of 200 quasi-random points in a two-dimensional space. The points appear to be evenly distributed across the whole space, providing better estimates and faster convergence in optimization tasks*

On the other hand, a space-filling design generated following a uniform random distribution may not provide even coverage of the whole space. The following code snippet generates 200 uniformly distributed points in the same two-dimensional unit hypercube:

Listing 5-6. Generating random samples following the uniform distribution

```
random_samples = torch.rand(200, 2)
xs_uniform = [x[0] for x in random_samples]
ys_uniform = [x[1] for x in random_samples]
plt.scatter(xs_uniform, ys_uniform)
```

Running this code snippet generates Figure 5-3, where empty regions are more prevalent compared to earlier. Leaving out these regions would thus provide a relatively insufficient coverage and suboptimal estimation.

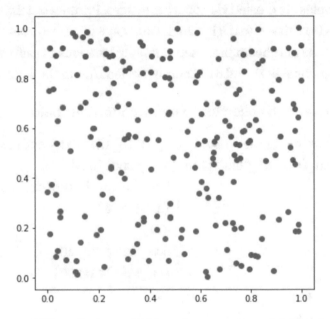

Figure 5-3. *A space-filling design of 200 points generated following a uniform distribution, where the coverage is notably less sufficient than that covered by the Sobol sequences*

After generating a good coverage using the Sobol sequences, the next step is to obtain the corresponding evaluations of the acquisition function. Since we have access to the analytic solution of EI, calculating these evaluations is cheap and straightforward. In Botorch, the Sobol sequences and corresponding values of the acquisition function are respectively stored in X_rnb and Y_rnb inside the gen_batch_initial_conditions() function.

The last step is to generate a set of initial conditions using the initialize_q_ batch() function, whose input arguments include X (the set of Sobol sequences), Y (corresponding EI values), and n (number of initial starts). Here, a heuristic is used to select the n initial conditions from the X Sobol sequences without replacement, where the probability of being selected is proportional to $e^{\eta Z}$, with η being an optional input temperature parameter and $Z = \dfrac{Y - \mu(Y)}{\sigma(Y)}$ representing the standardized EI value. In other words, a multinomial distribution is used to select the n initial conditions, where locations with high EI values are more likely to be selected, thus providing good initial starts to the multi-start optimization procedure.

Back to our running example. In the following code snippet, we generate 20 initial conditions in a six-dimensional unit hypercube using gen_batch_initial_conditions(), where 100 raw samples are created via Sobol sequences. Printing out the dimension of the returned result in Xinit shows that it is a 20x1x6 tensor, with 20 representing the number of initial conditions, 1 the number of proposed sampling locations in each BO outer iteration (i.e., the vanilla sequential BO), and 6 the number of features in the state space.

Listing 5-7. Generating a collection of initial conditions before optimization starts

```
from botorch.optim.optimize import gen_batch_initial_conditions
Xinit = gen_batch_initial_conditions(acq_function=EI,
                                     bounds=torch.tensor([[0.0] * 6,
                                     [1.0] * 6]),
                                     q=1,
                                     num_restarts=20,
                                     raw_samples=100)
>>> Xinit.shape
torch.Size([20, 1, 6])
```

When used in optimize_acqf(), gen_batch_initial_conditions() is used inside the wrapper function _gen_initial_conditions(), which uses another condition to decide if the acquisition function is knowledge gradient (KG), a topic to be covered in the next chapter. When there are no specific initial conditions provided, the _gen_initial_ conditions() function is used to generate the heuristic-based initial conditions. As a reference, the following code segment in optimize_acqf() handles the initial conditions before optimization starts. Also, note that the optimization routine starting from these initial starts will be performed in parallel to boost computational efficiency.

```
def _gen_initial_conditions() -> Tensor:
    ic_gen = (
        gen_one_shot_kg_initial_conditions
```

```
        if isinstance(acq_function, qKnowledgeGradient)
        else gen_batch_initial_conditions
    )
    batch_initial_conditions = ic_gen(
        acq_function=acq_function,
        bounds=bounds,
        q=q,
        num_restarts=num_restarts,
        raw_samples=raw_samples,
        fixed_features=fixed_features,
        options=options,
        inequality_constraints=inequality_constraints,
        equality_constraints=equality_constraints,
    )
    return batch_initial_conditions

if not initial_conditions_provided:
    batch_initial_conditions = _gen_initial_conditions()
```

The initial conditions generated via this function will get stored in the batch_initial_conditions variable. As a convention, the function used solely inside the high-level function starts with the underscore, as in the case of _gen_initial_conditions().

With the initial conditions generated based on the aforementioned heuristic, we will now enter into the core optimization phase using the gradient-based method with random restart, which is completed via the _optimize_batch_candidates() function, another built-in function in the source optimize_acqf() function.

As mentioned earlier, the actual optimization happens in gen_candidates_scipy(), which generates a set of optimal candidates based on the widely used scipy.optimize.minimize() function provided by SciPy. It starts with a set of initial candidates that serve as the starting points for optimization and are passed in the initial_conditions input argument. It will also accept additional core input arguments such as the acquisition function acquisition_function, lower bounds in lower_bounds, and upper bounds in upper_bounds.

Let us examine the core optimization routine closely. In the following code, we demonstrate the use of gen_candidates_scipy() in generating a set of optimal candidates, of which the candidate with the highest EI value is returned:

```
from botorch.generation.gen import gen_candidates_scipy
```

```
batch_candidates, batch_acq_values = gen_candidates_scipy(
            initial_conditions=Xinit,
            acquisition_function=EI,
            lower_bounds=torch.tensor([[0.0] * 6]),
            upper_bounds=torch.tensor([[1.0] * 6]),
        )
# extract the index of the best candidate
best = torch.argmax(batch_acq_values.view(-1), dim=0)
batch_candidates = batch_candidates[best]
```

Printing out the best candidate in batch_candidates shows the exact same candidate location as before:

```
>>> batch_candidates
tensor([[0.1715, 0.3180, 0.0816, 0.4204, 0.1985, 0.7727]])
```

Within gen_candidates_scipy(), the gradients of the loss (negative EI value) with respect to each initial condition are automatically calculated using the autograd feature of PyTorch. Specifically, the line torch.autograd.grad(loss, X) in f_np_wrapper() is used to perform automatic differentiation of the loss with respect to the input X. The following code snippet shows the definition of f_np_wrapper() for ease of reference, which returns EI evaluation in fval as well as the gradients in gradf at the candidate locations. Note that the losses at all candidate locations are added together to accelerate computation, where the total loss will be used to calculate the gradient with respect to each individual location.

Listing 5-8. Evaluating EI and its gradient at candidate locations

```
def f_np_wrapper(x: np.ndarray, f: Callable):
        """Given a torch callable, compute value + grad given a numpy
        array."""
        if np.isnan(x).any():
            raise RuntimeError(
                f"{np.isnan(x).sum()} elements of the {x.size}
                element array "
                f"`x` are NaN."
            )
```

```
X = (
    torch.from_numpy(x)
    .to(initial_conditions)
    .view(shapeX)
    .contiguous()
    .requires_grad_(True)
)
X_fix = fix_features(X, fixed_features=fixed_features)
loss = f(X_fix).sum()
# compute gradient w.r.t. the inputs (does not accumulate
in leaves)
gradf = _arrayify(torch.autograd.grad(loss, X)[0].contiguous().
view(-1))
if np.isnan(gradf).any():
    msg = (
        f"{np.isnan(gradf).sum()} elements of the {x.size}
        element "
        "gradient array `gradf` are NaN. This often indicates
        numerical issues."
    )
    if initial_conditions.dtype != torch.double:
        msg += " Consider using `dtype=torch.double`."
    raise RuntimeError(msg)
fval = loss.item()
return fval, gradf
```

With the EI value and the gradient available at any point across the search space, we can now utilize an off-the-shelf optimizer to perform the gradient-based optimization. A common routine is to call the "L-BFGS-B" method in the minimize() function, which is a quasi-Newton method that estimates the Hessian matrix (second derivative) of the objective function, in this case, the analytic EI. The following code snippet shows the part of gen_candidates_scipy() where the optimization happens, with the initial condition stored in x0, the objective function in f_np_wrapper, and the optimization results in res. For the full implementation details, the reader is encouraged to refer to the official BoTorch GitHub.

```
res = minimize(
        fun=f_np_wrapper,
        args=(f,),
        x0=x0,
        method=options.get("method", "SLSQP" if constraints else
        "L-BFGS-B"),
        jac=True,
        bounds=bounds,
        constraints=constraints,
        callback=options.get("callback", None),
        options={k: v for k, v in options.items() if k not in ["method",
        "callback"]},
    )
```

Up till now, we have examined the inner workings of optimizing over an analytic EI acquisition function. Note that while the number of proposals grows in the case of parallel BO, that is, q>1, the analytic solution is not readily available. In this case, the Monte Carlo (MC) version is often used to approximate the true value of the acquisition function at a specified input location. We refer to this as the MC acquisition function, as introduced in the next section.

Using MC Acquisition Function

When seeking the global optimum of the unknown black-box function, the acquisition functions are used as a heuristic to gauge the relative utility of each candidate sampling location. In this regard, there are two types of acquisition functions: those with analytic solutions and those without analytic expressions and whose value can only be approximated via Monte Carlo estimation for efficient computation. For example, an acquisition function that involves an expectation operator may be approximated using the average of many Monte Carlo samples simulated based on the updated posterior distribution.

Assuming we have observed a dataset $D_{1:n} = \{\mathbf{x}_i, y_i\}_{i=1}^n$ and would like to obtain the value of the acquisition function $\alpha(\mathbf{x}_{n+1})$ at an arbitrary next sampling location $\mathbf{x}_{n+1}$. Since the corresponding observation y_{n+1} is not yet revealed, our best strategy is to rely on the expected utility while taking into account the updated posterior distribution at

$\mathbf{x_{n+1}}$. Concretely, denote $\xi \sim P(f(\mathbf{x_{n+1}})|D_{1:n})$ as the random variable for the observation at $\mathbf{x_{n+1}}$. The value of the acquisition function $\alpha(\mathbf{x_{n+1}})$ would be conditioned on ξ, whose utility is denoted as $u(\xi)$. To integrate out the randomness in ξ, the expected utility is used as the value of the acquisition function, giving

$$\alpha(\mathbf{x_{n+1}}) = \mathbb{E}\left[u(\xi)|\xi \sim P\left(f(\mathbf{x_{n+1}})|D_{1:n}\right)\right]$$

where $P(f(\mathbf{x_{n+1}})|D_{1:n})$ is the posterior distribution of f at $\mathbf{x_{n+1}}$ given $D_{1:n}$ observed so far.

Evaluating the acquisition function $\alpha(\mathbf{x_{n+1}})$, however, requires taking the integral with respect to the posterior distribution, which is often analytically intractable and difficult to compute. This is even the case when proposing more than one sampling locations in parallel BO, where the analytic expressions generally do not exist. Instead, we would only be able to approximate $\alpha(\mathbf{x_{n+1}})$ by sampling a set of m putative Monte Carlo observations $\{\xi_i\}_{i=1}^{m}$ at $\mathbf{x_{n+1}}$, resulting in

$$\alpha(\mathbf{x_{n+1}}) \approx \frac{1}{m}\sum_{i=1}^{m}u(\xi_i)$$

where $\xi_i \sim P(f(\mathbf{x_{n+1}})|D_{1:n})$ is sampled based on the posterior distribution at $\mathbf{x_{n+1}}$. In the context of expected improvement, we would approximate the original closed-form EI via the following:

$$EI(\mathbf{x_{n+1}}) = \frac{1}{m}\sum_{i=1}^{m}\max\{\xi_i - f^*, 0\}$$

where f^* is the best observation collected so far in a noiseless observation model. $\xi_i \sim P(f(\mathbf{x_{n+1}})|D_{1:n})$ follows a normal distribution, where its mean and variance are denoted as $\mu(\mathbf{x_{n+1}})$ and $\sigma^2(\mathbf{x_{n+1}})$, respectively. In practical implementation, the reparameterization trick is often used to disentangle the randomness in ξ_i. Specifically, we can reexpress ξ_i as $\mu(\mathbf{x_{n+1}}) + L(\mathbf{x_{n+1}})\epsilon_i$, where $L(\mathbf{x_{n+1}})L(\mathbf{x_{n+1}})^T = \sigma^2(\mathbf{x_{n+1}})$ is the root decomposition of $\sigma^2(\mathbf{x_{n+1}})$, and $\epsilon_i \sim N(0, 1)$ follows a standard normal distribution.

In BoTorch, the MC EI is calculated by the qExpectedImprovement class, where the q represents the number of proposals in the parallel BO setting.

Using Monte Carlo Expected Improvement

The qExpectedImprovement class computes the batch EI (called qEI) value based on Monte Carlo simulation over multiple points, which are samples that follow Sobol sequences as described earlier. In this case, the Sobol sequences are sampled from the posterior normal distribution at a given sampling location, which are completed using the SobolQMCNormalSampler function.

In the following code snippet, we generate a total of 1024 random samples following Sobol sequences. First, following the reparameterization approach, a total of 1024 base samples are generated and stored in sampler. These samples then go through the scaling and shifting operations determined by the mean and variance of the posterior distribution, which is evaluated at the first observed location (i.e., train_x[0][None, :]) stored in posterior.

```
from botorch.sampling import SobolQMCNormalSampler
sampler = SobolQMCNormalSampler(1024, seed=1234)
posterior = model.posterior(train_x[0][None, :])
samples = sampler(posterior)
```

We can perform the sampling operation by passing posterior to sampler, where the results are printed as follows:

```
>>> samples
tensor([[[0.0506]],

        [[0.1386]],

        [[0.1737]],

        ...,

        [[0.1739]],

        [[0.1384]],

        [[0.0509]]], grad_fn=<UnsqueezeBackward0>)
```

The shape of samples also shows that there are a total of 1024 samples:

```
>>> samples.shape
torch.Size([1024, 1, 1])
```

With these random samples in place, the next step is to obtain the corresponding evaluations of the acquisition function so as to approximate its expectation. These evaluations are then averaged to produce the final approximated expected utility of the acquisition function at the specified sampling location, thus the name Monte Carlo acquisition function.

These computations are completed via the qExpectedImprovement class, which is designed to propose multiple points at the same time in the parallel BO framework. Specifically, the following four steps are performed:

- Sampling the joint posterior distribution over the q input points

- Evaluating the improvement over the current best value for each MC sample

- Maximizing over q points and selecting the best value

- Averaging over all the MC samples to produce the final estimate

Mathematically, $qEI(\mathbf{x})$ is defined as follows:

$$qEI(\mathbf{x}) = \mathbb{E}\left\{\max_{j=1,\ldots,q}\left[\max\left(\xi_j - f^*,0\right)\right]\right\}$$

where $\xi_i \sim P(f(\mathbf{x})|D)$. Using MC approximation together with the reparameterization trick, we can approximate $qEI(\mathbf{x})$ as follows:

$$qEI(\mathbf{x}) \approx \frac{1}{m}\sum_{i=1}^{m}\max_{j=1,\ldots,q}\left[\max\left(\mu(\mathbf{x})_j + \left(L(\mathbf{x})\epsilon_i\right)_j - f^*,0\right)\right]$$

where $\epsilon_i \sim N(0, 1)$. Let us look at an example on how an approximate qEI value can be calculated. In the following code snippet, we initialize a qExpectedImprovement instance by passing in the model object created earlier, the best-observed value so far in best_value, and the qMC sampler in sampler. The created instance in qEI could then be used to provide evaluation at any given sampling location. In this case, the evaluation is zero when passing in the first observed location, which makes sense since the marginal utility from any observed location should be none.

```
from botorch.acquisition import qExpectedImprovement
qEI = qExpectedImprovement(model, best_value, sampler)
>>> qEI(train_x[0][None, :])
tensor([0.], grad_fn=<MeanBackward1>)
```

Note that the [None, :] part essentially adds a batch dimension of 1 to the current tensor.

We can also plug in the best candidate location stored in batch_candidates and observe the qEI value, which is 0.0538:

```
>>> qEI(batch_candidates[None, :])
tensor([0.0538], grad_fn=<MeanBackward1>)
```

Let us compute the value using closed-form EI as follows:

```
>>> EI(batch_candidates[None, :])
tensor([0.0538], grad_fn=<MulBackward0>)
```

Within the qEI instance, the sequence of operations mentioned earlier is performed via the forward() function, which is included as follows for ease of reference:

Listing 5-9. Evaluating the qEI acquisition function

```
def forward(self, X: Tensor) -> Tensor:
    posterior = self.model.posterior(
        X=X, posterior_transform=self.posterior_transform
    )
    samples = self.sampler(posterior)
    obj = self.objective(samples, X=X)
    obj = (obj - self.best_f.unsqueeze(-1).to(obj)).clamp_min(0)
    q_ei = obj.max(dim=-1)[0].mean(dim=0)
    return q_ei
```

Next, let us apply this MC version of EI and seek the global optimum. In particular, we are interested in whether the MC acquisition function will return the same optimum location as the analytic counterpart. In the following code snippet, we first instantiate a normal sampler that produces 500 points of Sobol sequence, which will be used to estimate the value of MC EI at each proposed location via the MC_EI object. This acquisition function will then be passed into the same optimize_acqf() function that performs the core optimization procedure introduced earlier. We also set the seed of PyTorch to ensure reproducibility given the randomness in Monte Carlo simulations:

```
sampler = SobolQMCNormalSampler(num_samples=500, seed=0, resample=False)
# 500x100x1
MC_EI = qExpectedImprovement(
    model, best_f=best_value, sampler=sampler
)
torch.manual_seed(seed=0) # to keep the restart conditions the same
new_point_mc, _ = optimize_acqf(
    acq_function=MC_EI,
    bounds=torch.tensor([[0.0] * 6, [1.0] * 6]),
    q=1,
    num_restarts=20,
    raw_samples=100,
    options={},
)
```

Examining the result in new_point_mc shows that the location closely matches the one from analytic EI (saved in new_point_analytic):

```
>>> new_point_mc
tensor([[0.1715, 0.3181, 0.0816, 0.4204, 0.1985, 0.7726]])
```

We can also check the norm of the difference between these two vectors:

```
>>> torch.norm(new_point_mc - new_point_analytic)
tensor(0.0001)
```

Remember to check out the accompanying notebook at https://github.com/Apress/Bayesian-optimization/blob/main/Chapter_5.ipynb.

Summary

Many acquisition functions involve the expectation of some real-valued function of the model's output at a specific sampling location. Instead of directly evaluating the integral operation from the expectation operator, the Monte Carlo version of the acquisition function offers a convenient path to evaluate the value of the original acquisition function via MC simulations, thus avoiding integration and offering faster computation. Specifically, we covered the following:

- Optimizing over analytic expected improvement using BoTorch

- The multi-start optimization procedure used to seek the optimum solution via multiple paths in parallel, each starting from a random, heuristic-based initial condition

- The advantage of using Sobol sequences compared with uniform random sampling

- The Monte Carlo expected improvement acquisition function and the comparison with its counterpart

In the next chapter, we will discuss another acquisition function called knowledge gradient (KG) and the one-shot optimization technique when performing sequential search using KG.

## CHAPTER 6

Knowledge Gradient: Nested Optimization vs. One-Shot Learning

In the previous chapter, we learned the inner workings of the optimization procedure using BoTorch, highlighting the auto-differentiation mechanism and modular design of the framework. This paves the way for many new acquisition functions we can plug in and test. In this chapter, we will extend our toolkit of acquisition functions to the knowledge gradient (KG), a nonmyopic acquisition function that performs better than expected improvement (EI) in many cases.

Although empirically superior, calculating the KG value at an arbitrary location is nontrivial. The nonmyopic nature of the KG acquisition function increases the computational complexity since the analytic form is unavailable, and we need to perform a nested optimization for each KG evaluation; one needs to resort to the approximation method to obtain an evaluation of the KG value. Besides introducing the formulation of the KG acquisition function, this chapter also covers the mainstream approximation methods used to compute the KG value. In particular, we will illustrate the *one-shot KG* (OKG) formulation proposed in the BoTorch framework that can significantly accelerate the computation by converting the nested optimization problem into a deterministic optimization setting using techniques such as *sample average approximation* (SAA). We will also dive into the implementation of OKG in BoTorch, shedding light on the practice usage of this novel optimization technique.

© Peng Liu 2023
P. Liu, *Bayesian Optimization*, https://doi.org/10.1007/978-1-4842-9063-7_6

Introducing Knowledge Gradient

Knowledge gradient (KG) is a nonmyopic acquisition function that considers the impact of an additional putative observation on the marginal improvement of the expected utility, which takes the form of the posterior mean of the GP surrogate model based on a risk-neutral perspective. This means that a random variable is evaluated according to its expected value.

First introduced by Frazier et al. in their 2009 paper named "The Knowledge-Gradient Policy for Correlated Normal Beliefs," the KG acquisition function has enjoyed great popularity among the big family of acquisition functions due to its nonmyopic nature. In the paper, the authors interpret the amount of knowledge as the global reward (maximum posterior mean) based on a collected dataset $\mathcal{D}$ and the knowledge gradient acquisition function $\alpha_{KG}(\mathbf{x};\mathcal{D})$ as the expected increase in the knowledge after obtaining an evaluation at $\mathbf{x}$.

Let us start with the same setup, where we would like to maximize an unknown black-box function $f : \mathcal{X} \to \mathbb{R}$, where $\mathcal{X} \subset \mathbb{R}^d$ is a compact set. Assume we have collected a total of n pairs of location-value observations in $\mathcal{D}_n = \left\{(\mathbf{x}_i, y_i)\right\}_{i=1}^n$, and $y_i = f(\mathbf{x}_i) + \epsilon_i$ is the noise-perturbed observation model with $\epsilon_i \sim \mathcal{N}(0, \sigma^2)$. For an arbitrary location $\mathbf{x} \in \mathcal{X}$, we can rely on the closed-form expression for the posterior mean function and calculate its predictive mean function $\mu_{\mathcal{D}_n}(\mathbf{x}) = \mu(\mathbf{x};\mathcal{D}_n) = \mathbb{E}(y|\mathbf{x},\mathcal{D}_n)$.

Taking a similar approach as EI, we define the utility of the collected dataset $\mathcal{D}_n$ as the maximum value of the predictive mean function, that is, $u(\mathcal{D}_n) = \mu_{\mathcal{D}_n}^* = \max_{\mathbf{x} \in \mathcal{X}} \mu_{\mathcal{D}_n}(\mathbf{x})$. The utility increases if there is an increase in the maximum posterior mean, which is location independent. A shrewd reader may have immediately recognized the subtle difference compared with the utility function in EI, which uses the maximum observed value as the utility of the currently observed dataset $\mathcal{D}_n$ (assuming a noiseless model).

Another way to understand the role of the predictive mean function $\mu(\mathbf{x};\mathcal{D}_n)$ is that it serves as our estimate of the underlying objective function $f(\mathbf{x})$. That is, $\mu(\mathbf{x};\mathcal{D}_n)$ is the best average-case estimate of $f(\mathbf{x})$, and the corresponding predictive standard deviation function $\sigma(\mathbf{x};\mathcal{D}_n)$ quantifies the level of uncertainty at the particular location. As a result, the maximum $\mu_{\mathcal{D}_n}^*$ of the predictive mean function approximates the true global maximum $f^* = \max_{\mathbf{x} \in \mathcal{X}} f(\mathbf{x})$.

Suppose now we are to obtain an additional observation at $\mathbf{x}_{n+1}$ with the corresponding observation y_{n+1}. This additional observation would enlarge our training dataset to grow to $\mathcal{D}_{n+1} = \mathcal{D}_n \cup \left\{(\mathbf{x}_{n+1}, y_{n+1})\right\}$. We can quantitatively evaluate the value of this (and therefore arbitrarily any) location or equivalently the utility of the augmented dataset via the maximum of the new predictive function, that is, $u(\mathcal{D}_{n+1}) = \mu^*_{\mathcal{D}_{n+1}} = \max_{\mathbf{x} \in \mathcal{X}} \mu_{\mathcal{D}_{n+1}}(\mathbf{x})$. The improvement based on the additional pair $(\mathbf{x}_{n+1}, y_{n+1})$ could then be calculated as $\mu^*_{\mathcal{D}_{n+1}} - \mu^*_{\mathcal{D}_n}$.

We can then define the KG function as the expected marginal increase in utility between $\mathcal{D}_n$ and $\mathcal{D}_{n+1}$, where the nonmyopic nature is reflected as the expected outcome under various future simulations. Specifically, the KG function $\alpha_{KG}(\mathbf{x}; \mathcal{D}_n)$ at a candidate location $\mathbf{x}$ is defined as follows:

$$\alpha_{KG}(\mathbf{x}; \mathcal{D}_n) = \mathbb{E}_{p(y|\mathbf{x}, \mathcal{D}_n)}\left[\mu^*_{\mathcal{D}_{n+1}} - \mu^*_{\mathcal{D}_n}\right]$$

$$= \mathbb{E}_{p(y|\mathbf{x}, \mathcal{D}_n)}\left[\mu^*_{\mathcal{D}_{n+1}}\right] - \mu^*_{\mathcal{D}_n}$$

$$= \int\left[\max_{\mathbf{x}' \in \mathcal{X}} \mu_{\mathcal{D}_{n+1}}(\mathbf{x}')\right] p(y|\mathbf{x}, \mathcal{D}_n) dy - \max_{\mathbf{x}' \in \mathcal{X}} \mu_{\mathcal{D}_n}(\mathbf{x}')$$

Here, the expectation is taken with respect to the random variable y_{n+1} at a given location $\mathbf{x}_{n+1}$. Different realizations of y_{n+1} would result in different increases in the marginal gain in the utility function. By integrating out the randomness in y_{n+1}, we obtain the expected average-case marginal gain in utility. Such one-step lookahead formulation based on the global reward utility forms the KG acquisition function.

The global nature of KG also means that the last point to be evaluated may not necessarily be one of the locations previously evaluated; we are willing to commit to a new location upon the termination of the search process. For each new location, the integration effectively considers all possible values of observations under the predictive posterior mean function. Figure 6-1 summarizes the definition of the KG acquisition function.

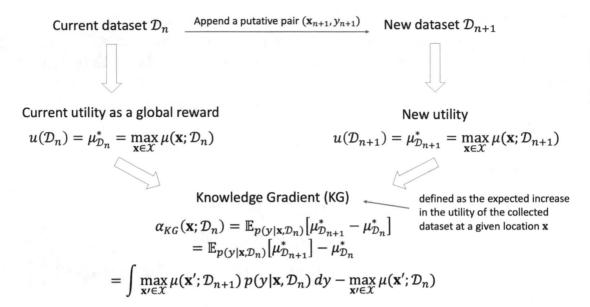

Figure 6-1. *Illustrating the knowledge gradient acquisition function*

Having formulated the KG acquisition function, we will look at its computation in the optimization procedure in the following section.

Monte Carlo Estimation

Given the expectation operator involved in the KG acquisition function, calculating the KG at each location is a nontrivial exercise since it is analytically intractable. However, we can use Monte Carlo (MC) simulations to approximate the KG value at a particular location. In this case, the KG acquisition function $\widehat{\alpha_{KG}}(\mathbf{x};\mathcal{D}_n)$ can be calculated using a total of M randomly generated samples from the current posterior GP model:

$$\widehat{\alpha_{KG}}(\mathbf{x};\mathcal{D}_n) = \frac{1}{M}\sum_{i=1}^{M}\left(u(\mathcal{D}_{n+1})^{(i)} - u(\mathcal{D}_n)\right)$$

$$= \frac{1}{M}\sum_{i=1}^{M}\left(\mu_{\mathcal{D}_{n+1}}^{*(i)} - \mu_{\mathcal{D}_n}^{*}\right)$$

where each ith putative sample $y^{(i)}$ is generated based on the corresponding GP posterior, that is, $y^{(i)} \sim p(y|\mathbf{x},\mathcal{D}_n)$. In addition, $\mu_{\mathcal{D}_{n+1}}^{*(i)} = \max\limits_{\mathbf{x}\in\mathcal{X}}\mu_{\mathcal{D}_{n+1}^{(i)}}(\mathbf{x})$, and $\mathcal{D}_{n+1}^{(i)} = \mathcal{D}_n \cup \left\{\left(\mathbf{x},y^{(i)}\right)\right\}$

is the simulation-augmented dataset after acquiring an additional data pair. As the number of Monte Carlo samples M increases, $\widehat{\alpha_{KG}}(\mathbf{x};\mathcal{D}_n)$ will approximate $\alpha_{KG}(\mathbf{x};\mathcal{D}_n)$ better. In the extreme case when $M \to \infty$, the approximation will be unbiased, that is, $\widehat{\alpha_{KG}}(\mathbf{x};\mathcal{D}_n)=\alpha_{KG}(\mathbf{x};\mathcal{D}_n)$.

Note that when locating the maximum posterior mean value, we can use the nonlinear optimization method such as L-BFGS-B to locate the global maximum along the posterior mean curve. L-BFGS-B is a second-order optimization algorithm we encountered previously when using the `minimize()` optimization routine provided by `SciPy`.

Figure 6-2 provides the general schematic of optimization using Monte Carlo simulations for certain acquisition functions, where MC integration in $\widehat{\alpha_{KG}}(\mathbf{x};\mathcal{D}_n)$ is used to approximate the expectation in $\alpha_{KG}(\mathbf{x};\mathcal{D}_n)$ using samples $\left\{y^{(i)}\right\}_{i=1}^{M}$ from the posterior distribution.

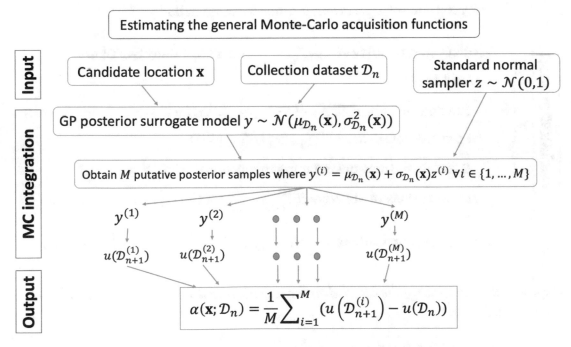

Figure 6-2. *Schematic of optimizing general Monte Carlo acquisition functions*

Let us look at the algorithm that can be used to evaluate the KG value at a given sampling location. As shown in Figure 6-3, we first calculate the maximum posterior mean value $\mu_{\mathcal{D}_n}^*$ based on the collected dataset $\mathcal{D}_n$ and then enter the main loop to calculate the approximate KG value based on simulations. In each iteration, we simulate a realization $y^{(i)}$ of a normally distributed random variable with mean $\mu_{\mathcal{D}_n}(\mathbf{x})$ and variance $\sigma^2_{\mathcal{D}_n}(\mathbf{x})$ at the location $\mathbf{x}$ under consideration. Equivalently, we can first generate a standard normal realization $z^{(i)} \sim \mathcal{N}(0,1)$ and apply the scale-shift transformation to get $y^{(i)} = \mu_{\mathcal{D}_n}(\mathbf{x}) + \sigma_{\mathcal{D}_n}(\mathbf{x})z^{(i)}$, a topic covered in Chapter 2. After obtaining the new simulation and appending it to the dataset to form $\mathcal{D}^{(i)}_{n+1}$, we can acquire a new maximum posterior mean value $\mu^{*(i)}_{\mathcal{D}_{n+1}}$, which can then be used to calculate the single improvement in utility, namely, $\left(\mu^{*(i)}_{\mathcal{D}_{n+1}} - \mu^*_{\mathcal{D}_n}\right)$. Upon completing the loop, the average improvement is returned as the approximate KG value,

that is, $\widehat{\alpha_{KG}}(\mathbf{x}; \mathcal{D}_n) = \dfrac{1}{M} \sum\limits_{i=1}^{M} \left(\mu^{*(i)}_{\mathcal{D}_{n+1}} - \mu^*_{\mathcal{D}_n}\right)$.

Obtaining the approximate Knowledge Gradient value via Monte Carlo simulations

Input

Observed dataset $\mathcal{D}_n$, current location $\mathbf{x}$, iteration budget M

Optimization loop

Calculate current utility value $\mu^*_{\mathcal{D}_n} = \max\limits_{\mathbf{x} \in \mathcal{X}} \mu(\mathbf{x}; \mathcal{D}_n)$ using L-BFGS-B

For i in 1 to M:

    Simulate an observation $y^{(i)} \sim \mathcal{N}(\mu_{\mathcal{D}_n}(\mathbf{x}), \sigma^2_{\mathcal{D}_n}(\mathbf{x}))$

    Form the new dataset $\mathcal{D}^{(i)}_{n+1} = \mathcal{D}_n \cup \{(\mathbf{x}, y^{(i)})\}$

    Calculate the new posterior mean function $\mu(\mathbf{x}'; \mathcal{D}^{(i)}_{n+1})$

    Calculate new utility value $\mu^{*(i)}_{\mathcal{D}_{n+1}} = \max\limits_{\mathbf{x}' \in \mathcal{X}} \mu(\mathbf{x}'; \mathcal{D}^{(i)}_{n+1})$ using L-BFGS-B

Output

Calculate the approximate KG value $\widehat{\alpha_{KG}}(\mathbf{x}; \mathcal{D}_n) = \dfrac{1}{M} \sum_{i=1}^{M} \left(\mu^{*(i)}_{\mathcal{D}_{n+1}} - \mu^*_{\mathcal{D}_n}\right)$

Figure 6-3. *Illustrating the algorithm for calculating the KG value using Monte Carlo simulations*

Therefore, using Monte Carlo simulation, we are essentially using the average difference in the maximum predictive mean between the augmented dataset $\mathcal{D}^{(M)}_{n+1}$ and the existing dataset $\mathcal{D}_n$ across all M simulations. This is also the inner optimization procedure which involves locating the maximum posterior mean, whereas the outer optimization procedure concerns locating the maximum KG value across the domain.

The computational cost, however, is extremely intensive in problems with higher dimensions. In the following section, we will zoom out and look at the overall optimization procedure using KG.

Optimizing Using Knowledge Gradient

Performing BO using KG is computationally intensive. For example, we first need to simulate M samples in order to calculate $\widehat{\alpha_{KG}}\left(\mathbf{x};\mathcal{D}_n\right)$ at a given location $\mathbf{x}$. Each simulated sample requires a maximization exercise to locate the maximum posterior mean based on the augmented dataset. To further add to the computational complexity of the algorithm, we need to evaluate different locations so as to identify the best sampling location with the highest KG value, that is, $\mathbf{x}_{n+1} = \underset{\mathbf{x} \in \mathcal{X}}{\operatorname{argmax}} \widehat{\alpha_{KG}}\left(\mathbf{x};\mathcal{D}_n\right)$.

To accelerate the optimization process, we could resort to first-order methods such as the stochastic gradient descent (SGD) introduced in the previous chapter. Under the multi-start scheme, we would initiate multiple SGD runs in parallel and observe the final converged locations, taking the one with the highest KG value as the next sampling location.

In order to apply the stochastic gradient descent method to move toward a local optimum point iteratively, we need to access the gradient of the (proxy) objective function with respect to the current location, that is, $\nabla_{\mathbf{x}} \alpha_{KG}\left(\mathbf{x};\mathcal{D}_n\right)$. Since $\alpha_{KG}\left(\mathbf{x};\mathcal{D}_n\right)$ is approximated by $\widehat{\alpha_{KG}}\left(\mathbf{x};\mathcal{D}_n\right)$, we can use SGD (more precisely, stochastic gradient ascent in the maximization setting) and calculate $\nabla_{\mathbf{x}} \widehat{\alpha_{KG}}\left(\mathbf{x};\mathcal{D}_n\right)$ to guide the search toward the local optimum. In other words, we would like to obtain the next evaluation location $\mathbf{x}_t + _1$ from the current location $\mathbf{x}_t$ using the following iteration:

$$\mathbf{x}_{t+1} = \mathbf{x}_t + \alpha_t \nabla_{\mathbf{x}_t} \alpha_{KG}\left(\mathbf{x}_t;\mathcal{D}_n\right)$$

Now let us look at the gradient term $\nabla_{\mathbf{x}_t} \alpha_{KG}\left(\mathbf{x}_t;\mathcal{D}_n\right)$. Plugging in the definition of $\alpha_{KG}\left(\mathbf{x}_t;\mathcal{D}_n\right)$ gives the following:

$$\nabla_{\mathbf{x}_t} \alpha_{KG}\left(\mathbf{x}_t;\mathcal{D}_n\right) = \nabla_{\mathbf{x}_t} \mathbb{E}_{p(y|\mathbf{x}_t,\mathcal{D}_n)}\left[\mu^*_{\mathcal{D}_{n+1}} - \mu^*_{\mathcal{D}_n}\right] = \nabla_{\mathbf{x}_t} \mathbb{E}_{p(y|\mathbf{x}_t,\mathcal{D}_n)}\left[\mu^*_{\mathcal{D}_{n+1}}\right]$$

Here, we removed the term $\mu^*_{\mathcal{D}_n}$ since the gradient does not depend on a fixed constant.

When taking the gradient of an expectation term, a useful technique called *infinitesimal perturbation analysis* can be applied here to ease the calculation, which essentially treats the gradient operator ∇ as a linear operator. Specifically, under sufficient regularity conditions about the underlying objective function, we can exchange the gradient and expectation operator, leading to the following:

$$\nabla_{\mathbf{x}_t} \alpha_{KG}\left(\mathbf{x}_t; \mathcal{D}_n\right) = \nabla_{\mathbf{x}_t} \mathbb{E}_{p(y|\mathbf{x}_t, \mathcal{D}_n)}\left[\mu^*_{\mathcal{D}_{n+1}}\right] = \mathbb{E}_{p(y|\mathbf{x}_t, \mathcal{D}_n)}\left[\nabla_{\mathbf{x}_t} \mu^*_{\mathcal{D}_{n+1}}\right]$$

With the expectation operator moved to the outer layer, we can apply the same Monte Carlo technique to sample a list of M gradients $\nabla_{\mathbf{x}_t} \mu^{*(i)}_{\mathcal{D}_{n+1}}$ for $i \in \{1, ..., M\}$ and take the average.

Now, the question is how to calculate a single gradient $\nabla_{\mathbf{x}_t} \mu^{*(i)}_{\mathcal{D}_{n+1}}$ for a given location $\mathbf{x}_t$ and simulated observation $y^{(i)}$. Recall that $\mu^{*(i)}_{\mathcal{D}_{n+1}}$ is the maximum posterior mean value after observing the putative pair $(\mathbf{x}_t, y^{(i)})$, that is:

$$\mu^{*(i)}_{\mathcal{D}_{n+1}} = \max_{\mathbf{x}' \in \mathcal{X}} \mu_{\mathcal{D}_{n+1}}\left(\mathbf{x}'\right)$$

In other words, we are calculating the gradient of a maximum over an infinite collection of posterior mean values located at different $\mathbf{x}'$. These posterior mean values are also functions of $\mathbf{x}_t$, the current location under consideration. Again, under sufficient regularity conditions about the underlying objective function, we can use the *envelope theorem* to simplify the gradient calculation. Specifically, we start by first identifying the maximum posterior mean among the infinite collection of posterior mean values, followed by differentiating the single maximum value (which is a function of $\mathbf{x}_t$) with respect to $\mathbf{x}_t$.

Denote $\widehat{\mathbf{x}_t^*} = \operatorname*{argmax}_{\mathbf{x}' \in \mathcal{X}} \mu_{\mathcal{D}_{n+1}}\left(\mathbf{x}'\right)$ as the location that maximizes the posterior mean function $\mu_{\mathcal{D}_{n+1}}\left(\mathbf{x}'\right)$. We can then keep $\widehat{\mathbf{x}_t^*}$ fixed and calculate the gradient of $\mu_{\mathcal{D}_{n+1}}\left(\widehat{\mathbf{x}_t^*}\right)$ with respect to $\mathbf{x}_t$, the current location under consideration, that is:

$$\nabla_{\mathbf{x}_t} \mu^*_{\mathcal{D}_{n+1}} = \nabla_{\mathbf{x}_t} \max_{\mathbf{x}' \in \mathcal{X}} \mu_{\mathcal{D}_{n+1}}\left(\mathbf{x}'\right) = \nabla_{\mathbf{x}_t} \mu_{\mathcal{D}_{n+1}}\left(\widehat{\mathbf{x}_t^*}\right)$$

We can then utilize the auto-differentiation capability in PyTorch to calculate the gradient with respect to the current location $\mathbf{x}_t$, after evaluating the posterior mean function at the maximizing location $\widehat{\mathbf{x}_t^*}$. Figure 6-4 summarizes the techniques involved in the derivation of gradient calculation.

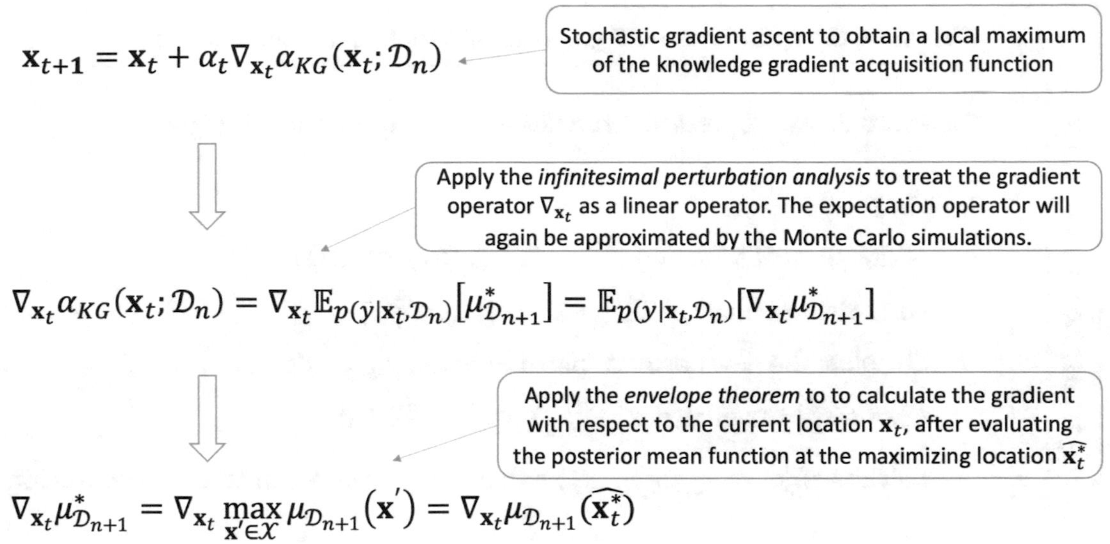

$$\mathbf{x}_{t+1} = \mathbf{x}_t + \alpha_t \nabla_{\mathbf{x}_t} \alpha_{KG}(\mathbf{x}_t; \mathcal{D}_n)$$

> Stochastic gradient ascent to obtain a local maximum of the knowledge gradient acquisition function

> Apply the *infinitesimal perturbation analysis* to treat the gradient operator $\nabla_{\mathbf{x}_t}$ as a linear operator. The expectation operator will again be approximated by the Monte Carlo simulations.

$$\nabla_{\mathbf{x}_t} \alpha_{KG}(\mathbf{x}_t; \mathcal{D}_n) = \nabla_{\mathbf{x}_t} \mathbb{E}_{p(y|\mathbf{x}_t, \mathcal{D}_n)}\left[\mu^*_{\mathcal{D}_{n+1}}\right] = \mathbb{E}_{p(y|\mathbf{x}_t, \mathcal{D}_n)}\left[\nabla_{\mathbf{x}_t} \mu^*_{\mathcal{D}_{n+1}}\right]$$

> Apply the *envelope theorem* to to calculate the gradient with respect to the current location $\mathbf{x}_t$, after evaluating the posterior mean function at the maximizing location $\widehat{\mathbf{x}^*_t}$

$$\nabla_{\mathbf{x}_t} \mu^*_{\mathcal{D}_{n+1}} = \nabla_{\mathbf{x}_t} \max_{\mathbf{x}' \in \mathcal{X}} \mu_{\mathcal{D}_{n+1}}(\mathbf{x}') = \nabla_{\mathbf{x}_t} \mu_{\mathcal{D}_{n+1}}(\widehat{\mathbf{x}^*_t})$$

Figure 6-4. *Illustrating the gradient calculation process for a given location*

Note that we still use the same Monte Carlo technique to handle the expectation operator via simulations. In Figure 6-5, we summarize the algorithm used to calculate the gradient element required in the SGD update, where we use $G_{\mathbf{x}_t}$ to denote the gradient of KG at the current running location $\mathbf{x}_t$. In this algorithm, we perform a total of J simulations to approximate the KG gradient. In each iteration, we first obtain a putative observation $y^{(j)} \sim \mathcal{N}\left(\mu_n(\mathbf{x}_t), \sigma_n^2(\mathbf{x}_t)\right)$ at the current running location $\mathbf{x}_t$ and form the new posterior mean function $\mu_{n+1}\left(\mathbf{x}'; \mathcal{D}_{n+1}^{(j)}\right)$. Next, we use the routine optimization protocol L-BFGS-B to locate the maximizer $\widehat{\mathbf{x}^*_t}$ and calculate a single gradient $G_{\mathbf{x}_t}^{(j)} = \nabla_{x_t} \mu_{\mathcal{D}_{n+1}}\left(\widehat{\mathbf{x}^*_t}\right)$ of the maximum posterior function value with respect to $\mathbf{x}_t$. Finally, we average multiple instances of $\left\{G_{\mathbf{x}_t}^{(j)}\right\}_{j=1}^J$ and return it as the approximate KG gradient $\nabla_{\mathbf{x}_t} \alpha_{KG}(\mathbf{x}_t; \mathcal{D}_n)$ for the current round of SGD update, that is, $\mathbf{x}_{t+1} = \mathbf{x}_t + \alpha_t \nabla_{\mathbf{x}_t} \alpha_{KG}(\mathbf{x}_t; \mathcal{D}_n)$. The averaging operation is theoretically guaranteed since the expected value of the average KG gradient is equal to the true gradient of the KG value, that is, $\mathbb{E}\left[G_{\mathbf{x}_t}\right] = \nabla_{\mathbf{x}_t} \alpha_{KG}(\mathbf{x}_t; \mathcal{D}_n)$.

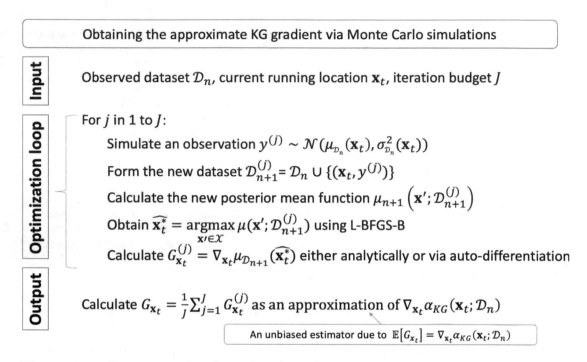

Figure 6-5. Illustrating the algorithm for calculating the gradient of KG using Monte Carlo simulations

Again, we can use the reparameterization trick, as mentioned in the previous chapter, to estimate the gradient $\nabla_{\mathbf{x}}\alpha_{KG}(\mathbf{x};\mathcal{D}_n)$ at a given location $\mathbf{x}$. Specifically, the random variable $y \sim \mathcal{N}\left(\mu_{\mathcal{D}_n}(\mathbf{x}),\sigma^2_{\mathcal{D}_n}(\mathbf{x})\right)$ to be sampled based on the predictive posterior distribution at location $\mathbf{x}$ given a collected dataset $\mathcal{D}_n$ can be expressed as $y = h_{\mathcal{D}_n}(\mathbf{x},z) = \mu_{\mathcal{D}_n}(\mathbf{x}) + \sigma_{\mathcal{D}_n}(\mathbf{x})z$, where $z \sim \mathcal{N}(0,1)$ and $h_{\mathcal{D}_n}(\mathbf{x},z)$ is a differentiable function with respect to $\mathbf{x}$. When $\alpha_{KG}(\mathbf{x};\mathcal{D}_n)$ is also differentiable with respect to $\mathbf{x}$, we have $\nabla_{\mathbf{x}}\alpha_{KG}(\mathbf{x};\mathcal{D}_n) = \nabla_h \alpha_{KG}(\mathbf{x};\mathcal{D}_n)\nabla_{\mathbf{x}}h_{\mathcal{D}_n}(\mathbf{x},z)$. Introducing the auxiliary random variable z via this differentiable and deterministic transformation thus helps in making the chain rule of differentiation explicit and less confounded by the parameters mean $\mu_{\mathcal{D}_n}(\mathbf{x})$ and variance $\sigma_{\mathcal{D}_n}(\mathbf{x})$ of the posterior distribution. See Figure 6-6 for an illustration on the reparameterization trick used to obtain an unbiased estimate of the gradient of the KG acquisition function.

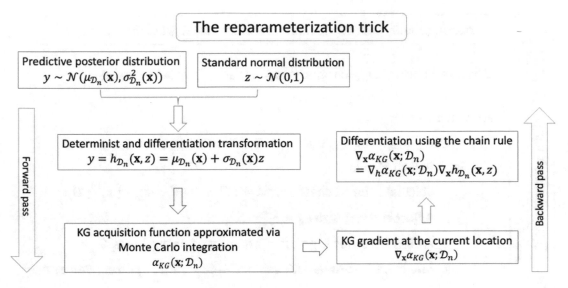

Figure 6-6. *Illustrating the reparameterization trick for the gradient calculation of the KG acquisition function*

Now we can shift to the outer optimization and look at the big picture: navigating toward the sampling location with a (locally) maximum KG value. As previously mentioned, we perform the multi-start stochastic gradient ascent procedure and select the best candidate with the highest KG value as the next sampling location.

Figure 6-7 provides the details on the algorithmic framework. Before optimization starts, we prepare the collected dataset $\mathcal{D}_n$, the total iteration budget R for the number of restarts and T for the number of gradient ascent updates, as well as the step size used for the gradient ascent update. We select a random point $\mathbf{x}_0^{(r)}$ as the starting point for each path of the stochastic gradient ascent update. Each starting point will evolve into a sequence of updates in $\mathbf{x}_t^{(r)}$ and converge to a local minimum at $\mathbf{x}_T^{(r)}$.

> **Locating the largest KG via multi-start stochastic gradient ascent**

Input

Observed dataset $\mathcal{D}_n$, iteration budget R and T, initial step size α

Optimization loop

For r in 1 to R:

    Pick a random starting point $\mathbf{x}_t^{(r)} \in \mathcal{X}^d$ at initial time point $t = 0$

    For t in 1 to $T - 1$:

        Calculate the stochastic gradient $G_{\mathbf{x}_t^{(r)}} \approx \nabla_{\mathbf{x}_t^{(r)}} \alpha_{KG}\left(\mathbf{x}_t^{(r)}; \mathcal{D}_n\right)$

        Adjust the step size $\alpha_t = \dfrac{\alpha}{\alpha + t}$

        Perform stochastic gradient ascent $\mathbf{x}_{t+1}^{(r)} = \mathbf{x}_t^{(r)} + \alpha_t G_{\mathbf{x}_t^{(r)}}$

    Calculate the approximate KG value $\widehat{\alpha_{KG}}\left(\mathbf{x}_T^{(r)}; \mathcal{D}_n\right)$ upon convergence

Output

Return the location with the largest KG value $\mathbf{x}_T^* = \underset{r \in \{1,\dots,R\}}{\operatorname{argmax}} \widehat{\alpha_{KG}}(\mathbf{x}_T^{(r)}; \mathcal{D}_n)$

Figure 6-7. *Illustrating the algorithm for locating the largest KG for sequential sampling using multi-start stochastic gradient ascent*

    To perform an ascent, we first calculate the approximate stochastic gradient $G_{\mathbf{x}_t^{(r)}} \approx \nabla_{\mathbf{x}_t^{(r)}} \alpha_{KG}\left(\mathbf{x}_t^{(r)}; \mathcal{D}_n\right)$ based on the Monte Carlo simulations covered earlier and perform the iterative stochastic gradient ascent update $\mathbf{x}_{t+1}^{(r)} = \mathbf{x}_t^{(r)} + \alpha_t G_{\mathbf{x}_t^{(r)}}$ to move to the next location. Note that the step size is made adaptive via $\alpha_t = \dfrac{\alpha}{\alpha + t}$, which reduces in magnitude as iteration proceeds. Each stochastic gradient ascent will run for a total of T iterations. Finally, we evaluate the approximate KG value $\widehat{\alpha_{KG}}\left(\mathbf{x}_T^{(r)}; \mathcal{D}_n\right)$ upon convergence of the ascent updates (set by the hyperparameter T) and return the location with the largest approximate KG value, that is, $\mathbf{x}_T^* = \operatorname{argmax} \widehat{\alpha_{KG}}\left(\mathbf{x}_T^{(r)}; \mathcal{D}_n\right)$.

    Due to its focus on improving the maximum posterior mean, the proposed next sampling location may not necessarily have a higher observation than previous observations. However, in the case of noisy observations, such nature in KG enables it to outperform the search strategy based on EI significantly.

    Note that the proposed suite of algorithms used to guide the sequential search process based on KG is much more efficient than the original direct computation. However, there could still be an improvement, especially when we need to propose multiple points at the same time, a topic under the theme of parallel Bayesian optimization. The corresponding state-of-the-art improvement over vanilla KG is called the one-shot KG (OKG), which will be introduced in the following section.

One-Shot Knowledge Gradient

When evaluating the KG acquisition function and its gradient estimator, we have iteratively used the Monte Carlo simulations to approximate the expectation operation. Instead of redrawing samples randomly each time, it turns out that there are several improvements we can make in terms of how and where we draw these samples.

One improvement is the quasi-Monte Carlo technique we introduced in the previous chapter. Instead of drawing the samples on a uniform scale, we use Sobol sequences to fill the sampling space more evenly, which allows us to converge faster and obtain more stable estimates. The simulated points will also have a low discrepancy in distance from one another, making it a more effective sampling approach than uniform sampling. This technique is referred to as the random quasi-Monte Carlo (RQMC) technique in the original BoTorch paper.

Another improvement introduced by the BoTorch paper is the sample average approximation (SAA), discussed in the next section.

Sample Average Approximation

The Monte Carlo simulation technique discussed so far requires redrawing the base samples $Z = \left\{z^{(i)}\right\}_{i=1}^{N}$ via $z \sim \mathcal{N}(0,1)$ for each round of the stochastic gradient descent update, where the repetitive drawing of new samples seems to be redundant. As a remedy, the sample average approximation technique avoids such redrawing and only draws a fixed set of base samples $Z = \left\{z^{(i)}\right\}_{i=1}^{N}$. This set of fixed base samples is used in each Monte Carlo evaluation throughout the optimization process. In other words, we fix a common set of random numbers, which turns the problem of optimizing the approximate KG function $\widehat{\alpha_{KG}}(\mathbf{x};\mathcal{D}_n)$ into a deterministic optimization exercise.

Note that since these base samples are independently and identically distributed (i.i.d.), the resulting estimator of the maximum KG value is theoretically guaranteed to converge to the true maximum, that is, $\widehat{\alpha}_N^* = \max_{\mathbf{x} \in \mathcal{X}} \widehat{\alpha_{KG}}(\mathbf{x};\mathcal{D}_n) \to \alpha^* = \max_{\mathbf{x} \in \mathcal{X}} \alpha_{KG}(\mathbf{x};\mathcal{D}_n)$. Such theoretical guarantee is backed by the rich literature on the approximation property of SAA, which also provides convenience to implementing MC-based approximation.

In addition, we can also use the same set of base samples to estimate the KG gradient value, although such a gradient estimator may be used for different purposes. Figure 6-8 illustrates the SAA process used in MC integration for both the KG value and its gradient estimator. The common random numbers $\left\{ z^{(i)} \right\}_{i=1}^{N}$ are usd to pass through a deterministic transformation $y = h_{\mathcal{D}_n}(\mathbf{x}, z) = \mu_{\mathcal{D}_n}(\mathbf{x}) + \sigma_{\mathcal{D}_n}(\mathbf{x}) z$ to produce a set of putative observations $\left\{ y^{(i)} \right\}_{i=1}^{N}$, which will then be used to calculate

$$\widehat{\alpha_{KG}}(\mathbf{x}; \mathcal{D}_n) = \frac{1}{N} \sum_{i=1}^{N} \left(\mu_{\mathcal{D}_{n+1}}^{*(i)} - \mu_{\mathcal{D}_n}^{*} \right) \text{ as an approximation of } \alpha_{KG}(x; \mathcal{D}_n) \text{ and}$$

$$G_{\mathbf{x}} = \frac{1}{N} \sum_{i=1}^{N} \nabla_{\mathbf{x}} \mu_{\mathcal{D}_{n+1}}\left(\widehat{\mathbf{x}}^{*} \right) \text{ as an approximation of } \nabla_{\mathbf{x}} \alpha_{KG}(\mathbf{x}; \mathcal{D}_n).$$

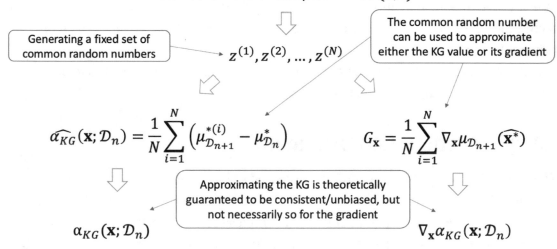

Figure 6-8. *The sample average approximation technique used in the Monte Carlo estimation for both the KG value and its gradient estimator using quasi-Newton methods for faster convergence speed and more robust optimization*

With a fixed set of base samples, we can now use second-order optimization routines such as quasi-Newton methods to achieve faster convergence speed and more robust optimization. This is because first-order methods such as stochastic gradient descent are subject to hyperparameter tuning, such as the learning rate, making it easier to achieve reliable optimization results with significant tuning efforts. When using second-order methods, however, such tuning is not necessary.

In the next section, we will discuss the one-shot formulation of KG using the SAA technique.

One-Shot Formulation of KG Using SAA

The optimization routine in seeking the next sampling location $\mathbf{x}^* = \operatorname*{argmax}_{\mathbf{x} \in \mathcal{X}} \alpha_{KG}(\mathbf{x}; \mathcal{D}_n)$ is based on a nested loop of optimization procedures. In the outer loop, we use the multi-start stochastic gradient ascent to locate the sampling location with the largest KG value, where both the KG value itself and its gradient are approximated via Monte Carlo simulations. In the inner loop, we seek the location with the maximum posterior mean, that is, $\mathbf{x}^* = \operatorname*{argmax}_{\mathbf{x} \in \mathcal{X}} \mu(\mathbf{x}; \mathcal{D}_{n+1})$, given the current sampling location $\mathbf{x}$ and simulated observation $y_{\mathcal{D}_n}(\mathbf{x})$. The inner loop can be performed either via the same first-order stochastic gradient ascent algorithm or the second-order L-BFGS-B procedure.

Let us now reexpress the KG value in more detail to facilitate the derivation of the one-shot KG formulation. Assume we take one more data pair $(\mathbf{x}, y_{\mathcal{D}_n}(\mathbf{x}))$ with $y_{\mathcal{D}_n}(\mathbf{x}) \sim \mathcal{N}(\mu_{\mathcal{D}_n}(\mathbf{x}), \sigma_{\mathcal{D}_n}(\mathbf{x}))$ and $\mathbb{E}[y_{\mathcal{D}_n}(\mathbf{x})] = \mu_{\mathcal{D}_n}(\mathbf{x})$, and we further form an updated (putative) dataset $\mathcal{D}_{n+1} = \mathcal{D}_n \cup \{(\mathbf{x}, y_{\mathcal{D}_n}(\mathbf{x}))\}$. The KG acquisition function can then be expressed as follows:

$$\alpha_{KG}(\mathbf{x}; \mathcal{D}_n) = \mathbb{E}_{p(y|\mathbf{x}, \mathcal{D}_n)}\left[\mu^*_{\mathcal{D}_{n+1}}\right] - \mu^*_{\mathcal{D}_n}$$

$$= \mathbb{E}_{p(y|\mathbf{x}, \mathcal{D}_n)}\left[\max_{\mathbf{x}' \in \mathcal{X}} \mu_{\mathcal{D}_{n+1}}(\mathbf{x}')\right] - \max_{\mathbf{x}' \in \mathcal{X}} \mu_{\mathcal{D}_n}(\mathbf{x}')$$

$$= \mathbb{E}_{p(y|\mathbf{x}, \mathcal{D}_n)}\left[\max_{\mathbf{x}' \in \mathcal{X}} \mathbb{E}_{p(y'|\mathbf{x}', \mathcal{D}_{n+1})}\left[y'_{\mathcal{D}_{n+1}}(\mathbf{x}')\right]\right] - \max_{\mathbf{x}' \in \mathcal{X}} \mathbb{E}\left[y_{\mathcal{D}_n}(\mathbf{x}') | \mathcal{D}_n\right]$$

$$= \mathbb{E}\left[\max_{\mathbf{x}' \in \mathcal{X}} \mathbb{E}\left[y'_{\mathcal{D}_{n+1}}(\mathbf{x}') | \mathcal{D}_{n+1}\right] | \mathcal{D}_n\right] - \max_{\mathbf{x}' \in \mathcal{X}} \mathbb{E}\left[y_{\mathcal{D}_n}(\mathbf{x}') | \mathcal{D}_n\right]$$

In other words, this expression essentially quantifies the expected marginal increase in the maximum posterior mean after sampling at location $\mathbf{x}$. When using SAA, we can avoid the nested optimization and use the fixed base samples to convert the original problem into a deterministic one. Specifically, after drawing a fixed set of base samples in $\{z^{(i)}\}_{i=1}^N$, we can approximate the KG acquisition function as follows:

$$\widehat{\alpha_{KG}}(\mathbf{x}; \mathcal{D}_n) = \frac{1}{N} \sum_{i=1}^N \max_{\mathbf{x}^{(i)} \in \mathcal{X}} \mathbb{E}\left[y_{\mathcal{D}_{n+1}}\left(\mathbf{x}^{(i)}\right) | \mathcal{D}_{n+1}^{(i)}\right] - \mu^*_{\mathcal{D}_n}$$

Since we would maximize the approximate KG function to obtain the next sampling position $\mathbf{x}_{n+1} = \widehat{\mathbf{x}}^* = \underset{\mathbf{x} \in \mathcal{X}}{\operatorname{argmax}} \widehat{\alpha}_{KG}(\mathbf{x}; \mathcal{D}_n)$, we can move the second inner maximization operator outside of the sample average when conditioned on the fixed set of base samples. In other words, we have

$$\max_{\mathbf{x} \in \mathcal{X}} \widehat{\alpha}_{KG}(\mathbf{x}; \mathcal{D}_n) = \max_{\mathbf{x} \in \mathcal{X}} \frac{1}{N} \sum_{i=1}^{N} \max_{\mathbf{x}^{(i)} \in \mathcal{X}} \mathbb{E}\left[y_{\mathcal{D}_{n+1}}\left(\mathbf{x}^{(i)}\right) | \mathcal{D}_{n+1}^{(i)} \right] - \mu_{\mathcal{D}_n}^*$$

$$= \max_{\mathbf{x} \in \mathcal{X}} \max_{\mathbf{x}^{(i)} \in \mathcal{X}} \frac{1}{N} \sum_{i=1}^{N} \mathbb{E}\left[y_{\mathcal{D}_{n+1}}\left(\mathbf{x}^{(i)}\right) | \mathcal{D}_{n+1}^{(i)} \right] - \mu_{\mathcal{D}_n}^*$$

$$= \max_{\mathbf{x}, \mathbf{x}' \in \mathcal{X}} \frac{1}{N} \sum_{i=1}^{N} \mathbb{E}\left[y_{\mathcal{D}_{n+1}}\left(\mathbf{x}^{(i)}\right) | \mathcal{D}_{n+1}^{(i)} \right] - \mu_{\mathcal{D}_n}^*$$

where $\mathbf{x}' := \left\{\mathbf{x}^{(i)}\right\}_{i=1}^{N} \in \mathcal{X}^N$ is a collection of N sampling points that represent the locations of the maximum posterior mean. These points are also called fantasy points in the BoTorch paper, used to present the next-stage solutions in the inner optimization loop. Using such reexpression, we have converted the original problem into an equivalent form with just one maximization operator, which requires us to solve an optimization problem over the current maximizing location $\mathbf{x}$ and next-stage fantasy points $\mathbf{x}'$.

Note that the only change in the current single optimization loop is that each loop needs to solve a higher-dimensional problem. That is, we need to output N more locations in each optimization. Since we are only working with a single optimization loop, such formulation is thus called the one-shot knowledge gradient (OKG). Figure 6-9 summarizes the formulation process of OKG.

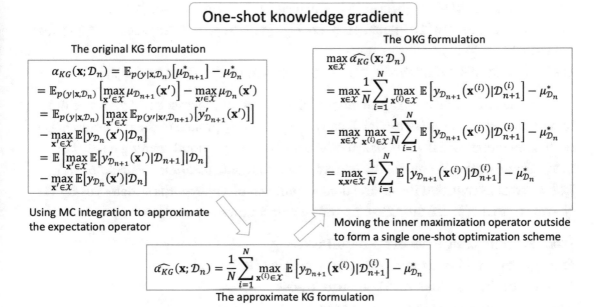

Figure 6-9. *Illustrating the one-shot knowledge gradient formulation*

In the following section, we will go through the implementation of OKG in BoTorch.

One-Shot KG in Practice

In this section, we will look at the implementation details of OKG in BoTorch based on a synthetic function of the form $f(x) = \sin(2\pi x_1) \cos(2\pi x_2)$, a two-dimensional function where all input features are confined to the hypercube $[0, 1]$. We also add a vector of normally distributed random noise to the simulated observations to simulate noise-perturbed observations. In addition, we standardize the noisy observations using the standardize() function in botorch.utils.transforms to create a vector of zero mean and unit variance. See the following code listing that implements these operations and creates 20 noisy observations after standardization:

Listing 6-1. Simulating 20 noisy observation

```
import os
import math
import torch
from botorch.utils.transforms import standardize
bounds = torch.stack([torch.zeros(2), torch.ones(2)])
```

```
train_X = bounds[0] + (bounds[1] - bounds[0]) * torch.rand(20, 2)
train_Y = torch.sin(2 * math.pi * train_X[:, [0]]) * torch.cos(2 * math.pi
* train_X[:, [1]])
train_Y = standardize(train_Y + 0.05 * torch.randn_like(train_Y))
```

Next, we will build a surrogate model using Gaussian processes and optimize the hyperparameters of the GP model. Following a similar approach to the previous chapter, we first instantiate a `SingleTaskGP` model based on the provided training dataset in `train_X` and `train_Y`, prepare the (closed-form) marginal likelihood calculator via `ExactMarginalLogLikelihood`, and finally optimize the GP hyperparameters using `fit_gpytorch_model()`. See the following code listing for these operations:

```
from botorch.models import SingleTaskGP
from gpytorch.mlls import ExactMarginalLogLikelihood
from botorch.fit import fit_gpytorch_model
model = SingleTaskGP(train_X, train_Y)
mll = ExactMarginalLogLikelihood(model.likelihood, model)
fit_gpytorch_model(mll);
```

Note that the semicolon is used here to suppress the output message after executing the code. Next, we will build a one-shot knowledge gradient instance to represent this particular acquisition function. In BoTorch, this class is implemented in `qKnowledgeGradient`, which generalizes to parallel Bayesian optimization setting where q represents the number of locations to be considered simultaneously. In addition, `qKnowledgeGradient` performs batch Bayesian optimization where each batch goes through the one-shot optimization process. See the following code listing on creating an instance for OKG learning later on:

```
from botorch.acquisition import qKnowledgeGradient

NUM_FANTASIES = 128
qKG = qKnowledgeGradient(model, num_fantasies=NUM_FANTASIES)
```

Here, we set the number of fantasy samples to 128, corresponding to the number of MC samples N for the outer expectation operator discussed earlier. More fantasy samples will have a better approximation of the true KG value, although at the expense of both RAM and wall time (running the whole script from start to end). As for the inner expectation, we can directly rely on the closed-form posterior mean or use another MC integration.

With the qKG instance created, we can use it to evaluate an arbitrary location across the search domain. Note that the one-shot formulation in OKG requires a total of 129 input locations per evaluation, including one location under consideration of the outer optimization and the remaining 128 fantasy locations used in MC integration.

The following code listing tests the OKG value of the first training location together with 128 fantasy locations sampled following Sobol sequences. We combine the original location with the 128 fantasy locations using the torch.cat() function, pass to qKG to evaluate the approximate KG value, and output the final content in the returned result by invoking the item() method:

```
from torch.quasirandom import SobolEngine
sobol_engine = SobolEngine(dimension=2, scramble=True, seed=8)
sobol_samples = sobol_engine.draw(NUM_FANTASIES)
>>> qKG(torch.cat((train_X[0].view(1,1),sobol_samples),0)).item()
0.19309169054031372
```

Let us look into the inner implementation of the qKnowledgeGradient class. By default, it uses the SobolQMCNormalSampler sampler to sample quasi-MC base samples using Sobol sequences. This is the same default sampler used in the qExpectedImprovement class discussed in the previous chapter and uses the reparameterization trick to generate posterior samples at given locations. In this case, we generate a total of num_fantasies base samples, all of which are fixed across the whole optimization process by setting resample=False, as shown in the following code snippet. Having fixed base samples is necessary when running deterministic optimization algorithm.

```
sampler = SobolQMCNormalSampler(
              num_samples=num_fantasies, resample=False, collapse_batch_
              dims=True
          )
```

Now let us look at the evaluation code when passing in a candidate location to the forward() function in qKnowledgeGradient. As shown in the following code snippet, we start by splitting the combined location vectors into the actual location in X_actual and fantasy locations in X_fantasies using the _split_fantasy_points() function:

Listing 6-2. Evaluating the qKG acquisition function

```python
def forward(self, X: Tensor) -> Tensor:
    # split fantasy location from the actual location under evaluation
    X_actual, X_fantasies = _split_fantasy_points(X=X, n_f=self.num_fantasies)

    # only concatenate X_pending into the X part after splitting
    if self.X_pending is not None:
        X_actual = torch.cat(
            [X_actual, match_batch_shape(self.X_pending, X_actual)], dim=-2
        )

    # construct the fantasy model of shape `num_fantasies x b`
    fantasy_model = self.model.fantasize(
        X=X_actual, sampler=self.sampler, observation_noise=True
    )

    # get the value function
    value_function = _get_value_function(
        model=fantasy_model,
        objective=self.objective,
        posterior_transform=self.posterior_transform,
        sampler=self.inner_sampler,
    )

    # make sure to propagate gradients to the fantasy model train inputs
    with settings.propagate_grads(True):
        values = value_function(X=X_fantasies)  # num_fantasies x b

    if self.current_value is not None:
        values = values - self.current_value

    # return average over the fantasy samples
    return values.mean(dim=0)
```

The following code snippet contains the _split_fantasy_points() function contained in BoTorch. Note that the X_fantasies variable goes through a reshaping operation to follow the shape of num_fantasies x batch_size x feature_dim to facilitate batch processing.

```
def _split_fantasy_points(X: Tensor, n_f: int) -> Tuple[Tensor, Tensor]:
    if n_f > X.size(-2):
        raise ValueError(
            f"n_f ({n_f}) must be less than the q-batch dimension of
            X ({X.size(-2)})"
        )
    split_sizes = [X.size(-2) - n_f, n_f]
    X_actual, X_fantasies = torch.split(X, split_sizes, dim=-2)
    X_fantasies = X_fantasies.permute(-2, *range(X_fantasies.
    dim() - 2), -1)
    X_fantasies = X_fantasies.unsqueeze(dim=-2)
    return X_actual, X_fantasies
```

Next, we would concatenate potential pending locations in X_pending to X_actual, if any. Then we construct a fantasy model based on the current location X_actual, where the fantasy model is used to create a new posterior GP model given simulated and possibly noise-perturbed observations. Specifically, the fantasy model is created by first computing the model posterior at X_actual, sampling from the posterior using the sampler provided to generate simulated observations, and finally conditioning the GP model on these new simulated observations to form an updated posterior distribution.

To see how the fantasy model works, let us look at an example of creating a fantasy model and generating fantasy evaluations. In the following code snippet, we create a sampler using the SobolQMCNormalSampler class by specifying num_samples=5, which creates five base samples following a Sobol sequence. These five base samples will be used together with the GP model (in the model object) to create artificial observations at a specified location (train_X[0] in this case). These operations are completed using the model's fantasize() method, which returns another GP model object in fantasy_model.

```
from botorch.sampling import SobolQMCNormalSampler
sampler = SobolQMCNormalSampler(num_samples=5, seed=1234)
fantasy_model = model.fantasize(train_X[0].view(1,-1), sampler,
observation_noise=True)
```

Both the model object and the fantasy_model object represent the GP model that is parameterized by the mean function and a covariance function conditioned on the observed/simulated dataset. In other words, we can supply an arbitrary sampling location and obtain the corresponding posterior mean and variance. For example, the following code snippet evaluates the posterior mean value at the location of the first training point using the `PosteriorMean()` function, a utility function used to evaluate the closed-form mean value from the posterior distribution:

```
from botorch.acquisition.analytic import PosteriorMean
>>> PosteriorMean(model)(train_X[0].view(1,-1))
tensor([-0.3386], grad_fn=<ViewBackward0>)
```

The result shows a scalar value of -0.4020 contained as a tensor object with the grad_fn attribute, which is the gradient at the specific location and will be used for gradient update later. We can also extract the scalar value by calling the `item()` function:

```
>>> PosteriorMean(model)(train_X[0].view(1,-1)).item()
-0.3386051654815674
```

We can apply the same operation to the fantasy model. As shown in the following code snippet, the fantasy model returns five different evaluations on the posterior mean due to five different fantasized posterior GP models:

```
>>> PosteriorMean(fantasy_model)(train_X[0].view(1,-1))
tensor([-0.4586, -0.2904, -0.2234, -0.3893, -0.3752], grad_fn=<ViewBackward0>)
```

Next, we construct the value function using `_get_value_function()`, which specifies the scoring function for a specific sampling location. In this case, we use the `PosteriorMean()` class to assign the posterior mean as the value to each sampling position. In other words, we use the posterior mean value to represent the absolute value of each sampling location. The following code snippet shows the definition of the `_get_value_function()` function from the BoTorch documentation:

Listing 6-3. Obtaining the surrogate value at any sampling location

```
def _get_value_function(
    model: Model,
    objective: Optional[MCAcquisitionObjective] = None,
    posterior_transform: Optional[PosteriorTransform] = None,
    sampler: Optional[MCSampler] = None,
    project: Optional[Callable[[Tensor], Tensor]] = None,
    valfunc_cls: Optional[Type[AcquisitionFunction]] = None,
    valfunc_argfac: Optional[Callable[[Model, Dict[str, Any]]]] = None,
) -> AcquisitionFunction:
```

```
r"""Construct value function (i.e. inner acquisition function)."""
if valfunc_cls is not None:
    common_kwargs: Dict[str, Any] = {
        "model": model,
        "posterior_transform": posterior_transform,
    }
    if issubclass(valfunc_cls, MCAcquisitionFunction):
        common_kwargs["sampler"] = sampler
        common_kwargs["objective"] = objective
    kwargs = valfunc_argfac(model=model) if valfunc_argfac is not
    None else {}
    base_value_function = valfunc_cls(**common_kwargs, **kwargs)
else:
    if objective is not None:
        base_value_function = qSimpleRegret(
            model=model,
            sampler=sampler,
            objective=objective,
            posterior_transform=posterior_transform,
        )
    else:
        base_value_function = PosteriorMean(
            model=model, posterior_transform=posterior_transform
        )

if project is None:
    return base_value_function
else:
    return ProjectedAcquisitionFunction(
        base_value_function=base_value_function,
        project=project,
    )
```

Note that we keep the setting to the simplest case; thus, the returned result is base_value_function=PosteriorMean(model=model, posterior_transform=posterior_transform). The value_function is then used to score the posterior mean across all the fantasy locations and return the average score as the final result of the qKG object.

Note that the calculation process of the posterior mean values using `value_function` is performed within the `propagate_grads` setting, which ensures that all calculations will backpropagate the gradient values in each tensor.

Up till now, we have looked at the inner workings of calculating the average posterior mean value as the approximate OKG value over a set of locations (current plus fantasy locations) instead of over a single current location in other acquisition functions. In the following section, we will look at the optimization process to identify the location set with the highest OKG value.

Optimizing the OKG Acquisition Function

Optimizing the OKG acquisition function follows the same modular procedure as the EI covered in the previous chapter. As shown in the following code snippet, we use the `optimize_acqf()` function to complete the optimization loop, where we specify `acq_function=qKG` to indicate the acquisition function used to measure the utility of each sampling location, `bounds=bounds` to indicate the search bound for each feature dimension, `q=2` to indicate the parallel BO that proposes two locations at the same time, `num_restarts=NUM_RESTARTS` to indicate the number of starting points in the multi-start optimization procedure, and `raw_samples=RAW_SAMPLES` to indicate the number of raw samples used for initialization across the whole search space:

Listing 6-4. Optimizing the OKG acquisition function

```
from botorch.optim import optimize_acqf
from botorch.utils.sampling import manual_seed

NUM_RESTARTS = 10
RAW_SAMPLES = 512

with manual_seed(1234):
    candidates, acq_value = optimize_acqf(
        acq_function=qKG,
        bounds=bounds,
        q=2,
        num_restarts=NUM_RESTARTS,
        raw_samples=RAW_SAMPLES,
    )
```

We can access the candidate locations and the corresponding OKG values as follows:

```
>>> candidates
tensor([[0.3940, 1.0000], [0.0950, 0.0000]])
>>> acq_value
tensor(2.0358)
```

The modular design in BoTorch makes it easy to plug in new acquisition functions and test the performance. Under the hood, the `optimize_acqf()` function has a separate handle on generating the initial conditions used for optimization, as indicated by the following code snippet inside the definition of the `optimize_acqf()` function from BoTorch:

```
ic_gen = (
            gen_one_shot_kg_initial_conditions
            if isinstance(acq_function, qKnowledgeGradient)
            else gen_batch_initial_conditions
        )
```

This snippet requires that the initial conditions are generated using the gen_one_shot_kg_initial_conditions() function when using the OKG acquisition function. The gen_one_shot_kg_initial_conditions() function generates a set of smart initial conditions using a combined strategy where some initial conditions are the maximizers of the simulated fantasy posterior mean.

The following code snippet shows the definition of the gen_one_shot_kg_initial_conditions() function in BoTorch. Here, we first get the total size of sampling locations to be selected simultaneously, where q is the number of parallel locations to be proposed. The definition of the get_augmented_q_batch_size() function is also given as follows, where we add the number of fantasy points to form the total dimension of sampling locations and save the result in q_aug:

Listing 6-5. Generating initial conditions for OKG

```
def gen_one_shot_kg_initial_conditions(
    acq_function: qKnowledgeGradient,
    bounds: Tensor,
    q: int,
    num_restarts: int,
```

```python
    raw_samples: int,
    fixed_features: Optional[Dict[int, float]] = None,
    options: Optional[Dict[str, Union[bool, float, int]]] = None,
    inequality_constraints: Optional[List[Tuple[Tensor, Tensor,
    float]]] = None,
    equality_constraints: Optional[List[Tuple[Tensor, Tensor,
    float]]] = None,
) -> Optional[Tensor]:
    options = options or {}
    frac_random: float = options.get("frac_random", 0.1)
    if not 0 < frac_random < 1:
        raise ValueError(
            f"frac_random must take on values in (0,1). Value: {frac_
            random}"
        )
    q_aug = acq_function.get_augmented_q_batch_size(q=q)

    ics = gen_batch_initial_conditions(
        acq_function=acq_function,
        bounds=bounds,
        q=q_aug,
        num_restarts=num_restarts,
        raw_samples=raw_samples,
        fixed_features=fixed_features,
        options=options,
        inequality_constraints=inequality_constraints,
        equality_constraints=equality_constraints,
    )

    # compute maximizer of the value function
    value_function = _get_value_function(
        model=acq_function.model,
        objective=acq_function.objective,
        posterior_transform=acq_function.posterior_transform,
        sampler=acq_function.inner_sampler,
        project=getattr(acq_function, "project", None),
```

```
)
from botorch.optim.optimize import optimize_acqf

fantasy_cands, fantasy_vals = optimize_acqf(
    acq_function=value_function,
    bounds=bounds,
    q=1,
    num_restarts=options.get("num_inner_restarts", 20),
    raw_samples=options.get("raw_inner_samples", 1024),
    fixed_features=fixed_features,
    return_best_only=False,
    inequality_constraints=inequality_constraints,
    equality_constraints=equality_constraints,
)

# sampling from the optimizers
n_value = int((1 - frac_random) * (q_aug - q))  # number of non-
random ICs
eta = options.get("eta", 2.0)
weights = torch.exp(eta * standardize(fantasy_vals))
idx = torch.multinomial(weights, num_restarts * n_value,
replacement=True)

# set the respective initial conditions to the sampled optimizers
ics[..., -n_value:, :] = fantasy_cands[idx, 0].view(num_restarts,
n_value, -1)
return ics

def get_augmented_q_batch_size(self, q: int) -> int:
    return q + self.num_fantasies
```

We start by optimizing the OKG acquisition function over a total of q_aug sampling locations following the parallel BO setting. Equating the nested optimization problem in the original KG formulation to the one-shot KG formulation and treating it as parallel BO is a key insight into understanding the OKG acquisition function. In this case, we adopt the same gen_batch_initial_conditions() function from the previous chapter on generating the heuristic-based initial conditions. This standard type of random initial condition is stored in ics.

Next, we use the same _get_value_function() function from OKG as the scoring function based on the posterior mean value. The scoring function is then used inside another optimization function in optimize_acqf() to locate the global maximum (with q=1) of the posterior mean among all fantasy models. In other words, the initial conditions are identified based on the maximizers of the posterior mean value of the fantasy models, which should intuitively be close to the maximizer of the current posterior mean. This type of nonrandom initial condition is stored in fantasy_cands.

Finally, we would choose a fraction of the initial conditions from the second type of nonrandom locations based on the specified input parameter frac_random. Specifically, the maximizing posterior mean values in fantasy_vals are used to weight a multinomial distribution after the softmax transformation, which is further used to sample from fantasy_cands and replace into ics. That is, the final initial conditions (sampling locations) are a mixture of random initializations and maximizing locations of the posterior mean of simulated fantasy models.

We can look at generating a set of initial conditions using the gen_one_shot_kg_ initial_conditions() function. In the following code snippet, we generate the initial conditions by setting options={"frac_random": 0.25}. By printing out the size of the result, we see that the dimension of the initial conditions follows num_restarts x (q+num_fantasies) x num_features.

```
from botorch.optim.initializers import gen_one_shot_kg_initial_conditions
Xinit = gen_one_shot_kg_initial_conditions(qKG, bounds, q=2, num_
restarts=10, raw_samples=512, options={"frac_random": 0.25})
>>> Xinit.size()
torch.Size([10, 130, 2])
```

Note that the maximizing OKG values we have calculated so far are based on the simulated posterior fantasy models. By definition, we also need to subtract the current maximum posterior mean to derive the expected increase in the marginal utility. In the following code snippet, we obtain the single maximum posterior mean based on the current model by setting acq_function=PosteriorMean(model) and q=1 in optimize_acqf():

```
from botorch.acquisition import PosteriorMean

NUM_RESTARTS = 20
RAW_SAMPLES = 2048
```

```
argmax_pmean, max_pmean = optimize_acqf(
    acq_function=PosteriorMean(model),
    bounds=bounds,
    q=1,
    num_restarts=NUM_RESTARTS,
    raw_samples=RAW_SAMPLES
)
```

We can observe the value of the maximum posterior mean as follows. The current maximum posterior mean value of 1.9548 is indeed lower than the new maximum posterior mean value of 2.0358 based on fantasized models:

```
>>> max_pmean
tensor(1.9548)
```

Now we can instantiate a proper version of OKG by setting current_value=max_ pmean in the initialization code as follows:

```
qKG_proper = qKnowledgeGradient(
    model,
    num_fantasies=NUM_FANTASIES,
    sampler=qKG.sampler,
    current_value=max_pmean,
)
```

Note that the current_value is used in the following snippet inside the definition of qKnowledgeGradient(), where the current maximum posterior mean is subtracted from the new fantasized maximum posterior mean:

```
if self.current_value is not None:
    values = values - self.current_value
```

Next, we apply the same optimization procedure to get the new candidate locations in candidates_proper and OKG value in acq_value_proper:

```
with manual_seed(1234):
    candidates_proper, acq_value_proper = optimize_acqf(
        acq_function=qKG_proper,
        bounds=bounds,
        q=2,
```

```
        num_restarts=NUM_RESTARTS,
        raw_samples=RAW_SAMPLES,
    )
>>> candidates_proper
tensor([[0.2070, 1.0000],
        [0.0874, 0.0122]])
>>> acq_value_proper
tensor(0.0107)
```

For the full scale of implementation covered, please visit the accompanying notebook at https://github.com/Apress/Bayesian-optimization/blob/main/Chapter_6.ipynb.

Summary

In this chapter, we have gone through the inner working of the one-shot knowledge gradient formulation and its implementation details in BoTorch. This recently proposed acquisition function enjoys attractive theoretical and practical properties. Specifically, we covered the following:

- The original KG formulation and its nested optimization procedure

- The reparameterization trick used to estimate Monte Carlo acquisition functions

- The Monte Carlo estimation technique used to approximate the expectation operation in calculating the KG acquisition function

- The multi-start stochastic gradient ascent algorithm with the approximate KG gradient estimator to locate the maximum KG value

- Deriving the one-shot KG acquisition function via sample average approximation

- Implementation of the OKG optimization procedure in BoTorch

In the next chapter, we will cover a case study that fine-tunes the learning rate when training a deep convolutional neural network.

## CHAPTER 7

Case Study: Tuning CNN Learning Rate with BoTorch

By now, we have established a good foundation regarding the theoretical inner workings of a typical Bayesian optimization process: a surrogate function that approximates the underlying true function and gets updated as new data arrives and an acquisition function that guides the sequential search under uncertainty. We have covered popular choices of acquisition function, including expected improvement (EI, with its closed-form expression derived in Chapter 3), the more general Monte Carlo acquisition functions such as knowledge gradient (KG, computed via multi-start stochastic ascent in Chapter 6), and their parallel versions (qEI and qKG) and multi-step lookahead extensions (both accelerated via one-shot optimization, also in Chapter 6). This foundation carries us a long way, as we will demonstrate via two case studies: a simulated case that seeks the global maximum of the six-dimensional Hartmann function and an empirical case that finds the optimal learning rate of a convolutional neural network (CNN) used to classify the widely used MNIST dataset.

We wish to highlight another important purpose of this chapter: to showcase the full Bayesian optimization ecosystem using BoTorch and provide a template for users interested in applying this principled search policy in their respective problem domains. Although more integrated and user-friendly options exist out there, such as the Ax platform, we will stay with the "build from scratch" style of this book, which also gives users the maximal level of control (and appreciation) of the ins and outs.

© Peng Liu 2023
P. Liu, *Bayesian Optimization*, https://doi.org/10.1007/978-1-4842-9063-7_7

Seeking Global Optimum of Hartmann

We encountered the Hartmann function in Chapter 5. There, we demonstrated a single step of search using qEI. We will now extend to multiple steps of search, compare and contrast multiple acquisition functions, and measure the quality of the search as the optimization proceeds.

To refresh our memory, recall that the Hartmann function is a six-dimensional function that has six local minima and one global minimum. Two flavors make it a challenging task to seek the global minimum of this function: a high-dimensional search space and a highly nonconvex function surface. The function takes the following form:

$$f(\mathbf{x}) = -\sum_{i=1}^{4} \alpha_i \exp\left(-\sum_{j=1}^{6} A_{ij} \left(x_j - P_{ij} \right)^2 \right)$$

where α_i, A_{ij}, and P_{ij} are all constants. The search space is a six-dimensional unit hypercube, with each action $x_j \in (0, 1)$ for all $j = 1, \ldots, 6$. The global minimum is obtained at $\mathbf{x}^* = (0.20169, 0.150011, 0.476874, 0.275332, 0.311652, 0.6573)$, with $f(\mathbf{x}^*) = -3.32237$.

Let us start by importing a few common packages as follows. Note that it is always a good practice to configure the random seed of all the three common packages (`random`, `numpy`, and `torch`) *before* we start to write any code. Here, we set the random seed to 8 for all three packages:

```
import os
import math
import torch
import random
import numpy as np
from matplotlib import pyplot as plt
%matplotlib inline

SEED = 8
random.seed(SEED)
np.random.seed(SEED)
torch.manual_seed(SEED)
```

We would also like to configure two elements: the computing device and the tensor objects' data type. In the following code snippet, we specify the device to be GPU if it is available; having a GPU will significantly accelerate the computation in both BO loop and training neural networks. We additionally assign the data type to be double for more precise computation:

```
device = torch.device("cuda" if torch.cuda.is_available() else "cpu")
dtype = torch.double
```

We can load the Hartmann function as our unknown objective function and negate it to fit the maximization setting as before:

```
# unknown objective function
from botorch.test_functions import Hartmann

neg_hartmann6 = Hartmann(negate=True)
```

Now we can generate initial conditions in the form of a set of randomly selected input locations and the corresponding noise-corrupted observations.

Generating Initial Conditions

We would like to define a function that gives us a set of initial conditions based on the specified number of observations. Each condition comes in pairs: the input location and the corresponding noise-disturbed observation value. The following code snippet defines such a function named generate_initial_data(), where we specify a noise level of 0.1 as the standard deviation of the observation model:

Listing 7-1. Generating initial conditions

```
# generate initial dataset
NOISE_SE = 0.1

def generate_initial_data(n=10):
    # generate random initial locations
    train_x = torch.rand(n, 6, device=device, dtype=dtype)
    # obtain the exact value of the objective function and add output
    dimension
    exact_obj = neg_hartmann6(train_x).unsqueeze(-1)
    # add Gaussian noise to the observation model
    train_y = exact_obj + NOISE_SE * torch.randn_like(exact_obj)
```

```
# get the current best observed value, i.e., utility of the
available dataset
best_observed_value = train_y.max().item()
return train_x, train_y, best_observed_value
```

Let us test this function and generate five initial conditions, as shown in the following code snippet. The result shows that both the input location variable train_x and observation variable train_y live in the GPU (device='cuda:0') and assume a data type of float (torch.float64). The best-observed value is 0.83, which we seek to improve upon in the follow-up sequential search.

```
>>>train_x, train_y, best_observed_value = generate_initial_data(n=5)
>>>print(train_x)
>>>print(train_y)
>>>print(best_observed_value)
tensor([[0.7109, 0.0106, 0.2621, 0.0697, 0.2866, 0.5305],
        [0.0386, 0.5117, 0.8321, 0.8719, 0.2094, 0.1702],
        [0.7290, 0.5147, 0.9309, 0.8339, 0.1056, 0.6658],
        [0.6747, 0.4133, 0.5129, 0.3619, 0.6991, 0.8187],
        [0.5215, 0.8842, 0.3067, 0.2046, 0.1769, 0.1818]], device='cuda:0',
       dtype=torch.float64)
tensor([[ 0.8309],
        [ 0.0752],
        [-0.0698],
        [ 0.0430],
        [ 0.5490]], device='cuda:0', dtype=torch.float64)
0.8308929843287978
```

Now we use these initial conditions to obtain the posterior distributions of the GP model.

Updating GP Posterior

We are concerned with two objects in the language of BoTorch regarding the GP surrogate model: the GP model itself and the marginal log-likelihood given the set of collected observations. We define the following function to digest the dataset of initial conditions and then return these two objects:

```
# initialize GP model
from botorch.models import SingleTaskGP
from gpytorch.mlls import ExactMarginalLogLikelihood

def initialize_model(train_x, train_y):
    # create a single-task exact GP model instance
    # use a GP prior with Matern kernel and constant mean function
    by default
    model = SingleTaskGP(train_X=train_x, train_Y=train_y)
    mll = ExactMarginalLogLikelihood(model.likelihood, model)

    return mll, model
```

Here, we use the `SingleTaskGP` class to build a single-output GP model and the `ExactMarginalLogLikelihood()` function to obtain the marginal log-likelihood based on the closed-form expression. We can then use this function to create the two variables that host the GP model and its marginal log-likelihood, respectively, in addition to printing out the value of the hyperparameters for the GP model:

```
mll, model = initialize_model(train_x, train_y)
>>> list(model.named_hyperparameters())
[('likelihood.noise_covar.raw_noise', Parameter containing:
  tensor([2.0000], device='cuda:0', dtype=torch.float64, requires_
  grad=True)),
 ('mean_module.raw_constant', Parameter containing:
  tensor(0., device='cuda:0', dtype=torch.float64, requires_grad=True)),
 ('covar_module.raw_outputscale', Parameter containing:
  tensor(0., device='cuda:0', dtype=torch.float64, requires_grad=True)),
 ('covar_module.base_kernel.raw_lengthscale', Parameter containing:
  tensor([[0., 0., 0., 0., 0., 0.]], device='cuda:0', dtype=torch.float64,
          requires_grad=True))]
```

Optimizing the GP model's hyperparameters (kernel parameters and noise variance) is completed using the `fit_gpytorch_mll()` function. However, since all the codes are written and run in Google Colab, we found that this step requires sending the marginal log-likelihood object `mll` to the CPU before calling the `fit_gpytorch_mll()` function. As shown in the following code snippet, we perform the movement via `mll.cpu()` and then move it back to the GPU via `mll.to(train_x)`, which converts `mll` to the same data type and location as `train_x`:

Listing 7-2. Optimizing GP hyperparameters

```
# fit GP hyperparameters
from botorch.fit import fit_gpytorch_mll
# fit hyperparameters (kernel parameters and noise variance) of a
GPyTorch model
fit_gpytorch_mll(mll.cpu());
mll = mll.to(train_x)
model = model.to(train_x)
```

We can also verify if the model object is indeed in the GPU:

```
>>> print(next(model.parameters()).is_cuda)
True
```

The posterior GP model with optimized hyperparameters now looks different:

```
>>> list(model.named_hyperparameters())
[('likelihood.noise_covar.raw_noise', Parameter containing:
  tensor([0.0201], device='cuda:0', dtype=torch.float64, requires_
  grad=True)),
 ('mean_module.raw_constant', Parameter containing:
  tensor(0.2875, device='cuda:0', dtype=torch.float64, requires_grad=True)),
 ('covar_module.raw_outputscale', Parameter containing:
  tensor(-1.5489, device='cuda:0', dtype=torch.float64, requires_
  grad=True)),
 ('covar_module.base_kernel.raw_lengthscale', Parameter containing:
  tensor([[-0.8988, -0.9278, -0.9508, -0.9579, -0.9429, -0.9305]],
         device='cuda:0', dtype=torch.float64, requires_grad=True))]
```

Next, we will define an acquisition function to perform the sequential search.

Creating a Monte Carlo Acquisition Function

We demonstrate a single step of sequential search using the qEI acquisition function, which is implemented in the qExpectedImprovement class in botorch.acquisition. Since qEI proposes multiple locations at the same time and its closed-form expression is not available, we would need to invoke Monte Carlo techniques to approximate the expectation operator required by the definition of qEI.

Toward this end, we use the SobolQMCNormalSampler class to create a sampler instance that samples Sobol sequences for a more stable approximation, sampling 256 points at each Monte Carlo estimation as shown in the following code snippet:

```
# define a QMC sampler to sample the acquisition function
from botorch.sampling.normal import SobolQMCNormalSampler

MC_SAMPLES = 256
qmc_sampler = SobolQMCNormalSampler(sample_shape=torch.Size([MC_SAMPLES]))
```

The qEI acquisition function requires three input arguments: the GP model object, the best-observed value so far, and the quasi-Monte Carlo sampler:

```
# use qEI as the MC acquisition function
from botorch.acquisition import qExpectedImprovement

qEI = qExpectedImprovement(model, best_observed_value, qmc_sampler)
```

Now let us define a function that takes the specified acquisition function as input, performs the inner optimization to obtain the maximizer of the acquisition function using optimize_acqf(), and returns the proposed location and noise-corrupted observation as the output.

In the following code snippet, we define the bound variable to set the upper and lower bounds to be the unit hypercube with the preassigned computing device and data type. We also use BATCH_SIZE to indicate the number of parallel candidate locations to be generated in each BO iteration, NUM_RESTARTS to indicate the number of starting points used in the multi-start optimization procedure, and RAW_SAMPLES to denote the number of samples used for initialization when optimizing the MC acquisition function:

Listing 7-3. Optimizing the acquisition function

```
# optimize and get new observation
from botorch.optim import optimize_acqf

# 6d unit hypercube
bounds = torch.tensor([[0.0] * 6, [1.0] * 6], device=device, dtype=dtype)
# parallel candidate locations generated in each iteration
BATCH_SIZE = 3
# number of starting points for multistart optimization
NUM_RESTARTS = 10
```

```python
# number of samples for initialization
RAW_SAMPLES = 512

def optimize_acqf_and_get_observation(acq_func):
    """Optimizes the acquisition function, and returns a new candidate and
    a noisy observation."""
    # optimize
    candidates, _ = optimize_acqf(
        acq_function=acq_func,
        bounds=bounds,
        q=BATCH_SIZE,
        num_restarts=NUM_RESTARTS,
        raw_samples=RAW_SAMPLES,  # used for initialization heuristic
        options={"batch_limit": 5, "maxiter": 200},
    )

    # observe new values
    new_x = candidates.detach()
    exact_obj = neg_hartmann6(new_x).unsqueeze(-1)  # add output dimension
    new_y = exact_obj + NOISE_SE * torch.randn_like(exact_obj)

    return new_x, new_y
```

Now let us test this function and observe the next batch of three sampling locations proposed and observations from the observation model. The result shows that both new_x and new_y contain three elements, all living in the GPU and assuming a floating data type:

```python
new_x, new_y = optimize_acqf_and_get_observation(qEI)
>>> print(new_x)
>>> print(new_y)
tensor([[0.9381, 0.0000, 0.2803, 0.0094, 0.3038, 0.4935],
        [0.6015, 0.0000, 0.1978, 0.0167, 0.1881, 0.6974],
        [0.5998, 0.0529, 0.2400, 0.1048, 0.2779, 0.3232]], device='cuda:0',
      dtype=torch.float64)
tensor([[0.2815],
        [0.6528],
        [0.5882]], device='cuda:0', dtype=torch.float64)
```

We now introduce the main course and move into the full BO loop.

The Full BO Loop

The full BO loop refers to an iterative sequential search up to a fixed number of iterations. To compare multiple search strategies, we use three policies: qEI, qKG, and the random search policy that picks three random points at each step. We would like to maintain three lists to record the respective best-observed value after each BO iteration. Moreover, we intend to run the full BO loop several times to check the average performance and its variance across multiple random starts.

The following code snippet provides the implementation of the full BO loop. To start with, we set N_TRIALS = 3 to repeat all experiments three times and assess the variation in the resulting performance, verbose = True to print experiment diagnostics, and N_BATCH = 20 to indicate the number of iterations in the outer BO loop, and create three placeholders (best_observed_all_qei, best_observed_all_qkg, and best_random_all) to hold the experiment results of all three search policies, respectively:

Listing 7-4. The full BO loop on global optimization

```
import time
from botorch.acquisition import qKnowledgeGradient

# number of runs to assess std of different BO loops
N_TRIALS = 3
# indicator to print diagnostics
verbose = True
# number of steps in the outer BO loop
N_BATCH = 20
best_observed_all_qei, best_observed_all_qkg, best_random_all = [], [], []

# average over multiple trials
for trial in range(1, N_TRIALS + 1):

    print(f"\nTrial {trial:>2} of {N_TRIALS} ", end="")
    best_observed_qei, best_observed_qkg, best_random = [], [], []

    # call helper functions to generate initial training data and
    # initialize model
    train_x_qei, train_y_qei, best_observed_value_qei = generate_initial_
    data(n=10)
    mll_qei, model_qei = initialize_model(train_x_qei, train_y_qei)
```

```
train_x_qkg, train_y_qkg = train_x_qei, train_y_qei
best_observed_value_qkg = best_observed_value_qei
mll_qkg, model_qkg = initialize_model(train_x_qkg, train_y_qkg)

best_observed_qei.append(best_observed_value_qei)
best_observed_qkg.append(best_observed_value_qkg)
best_random.append(best_observed_value_qei)

# run N_BATCH rounds of BayesOpt after the initial random batch
for iteration in range(1, N_BATCH + 1):

    t0 = time.monotonic()

    # fit the models
    fit_gpytorch_mll(mll_qei.cpu());
    mll_qei = mll_qei.to(train_x)
    model_qei = model_qei.to(train_x)

    fit_gpytorch_mll(mll_qkg.cpu());
    mll_qkg = mll_qkg.to(train_x)
    model_qkg = model_qkg.to(train_x)

    # define the qEI acquisition function using a QMC sampler
    qmc_sampler = SobolQMCNormalSampler(sample_shape=torch.Size([MC_
    SAMPLES]))

    # for best_f, we use the best observed noisy values as an
    approximation
    qEI = qExpectedImprovement(
        model = model_qei,
        best_f = train_y.max(),
        sampler = qmc_sampler
    )

    qKG = qKnowledgeGradient(model_qkg, num_fantasies=MC_SAMPLES)

    # optimize and get new observation
    new_x_qei, new_y_qei = optimize_acqf_and_get_observation(qEI)
    new_x_qkg, new_y_qkg = optimize_acqf_and_get_observation(qKG)

    # update training points
```

```python
    train_x_qei = torch.cat([train_x_qei, new_x_qei])
    train_y_qei = torch.cat([train_y_qei, new_y_qei])
    train_x_qkg = torch.cat([train_x_qkg, new_x_qkg])
    train_y_qkg = torch.cat([train_y_qkg, new_y_qkg])

    # update progress
    best_random = update_random_observations(best_random)
    best_value_qei = neg_hartmann6(train_x_qei).max().item()
    best_value_qkg = neg_hartmann6(train_x_qkg).max().item()
    best_observed_qei.append(best_value_qei)
    best_observed_qkg.append(best_value_qkg)

    # reinitialize the models so they are ready for fitting on next
    iteration
    mll_qei, model_qei = initialize_model(
        train_x_qei,
        train_y_qei
    )
    mll_qkg, model_qkg = initialize_model(
        train_x_qkg,
        train_y_qkg
    )

    t1 = time.monotonic()

    if verbose:
        print(
            f"\nBatch {iteration:>2}: best_value (random, qEI, qKG) = "
            f"({max(best_random):>4.2f}, {best_value_qei:>4.2f}, {best_
            value_qkg:>4.2f}),"
            f"time = {t1-t0:>4.2f}.", end=""
        )
    else:
        print(".", end="")

best_observed_all_qei.append(best_observed_qei)
best_observed_all_qkg.append(best_observed_qkg)
best_random_all.append(best_random)
```

Note that the basic BO procedure stays the same in this seemingly long-running code listing. For each trial, we would generate a set of initial conditions using `generate_initial_data()`, initialize the GP model using `initialize_model()`, optimize its hyperparameters using `fit_gpytorch_mll()`, employ `optimize_acqf_and_get_observation()` to obtain the next batch of sampling locations and noise-perturbed observation for the respective search policy, and finally update the results in the master lists.

The random search policy is a special and simple one; we choose a random set of locations, collect the noisy observation, and record the current best candidate. The following function provides the update rule:

Listing 7-5. The random policy

```
def update_random_observations(best_random):
    """Simulates a random policy by drawing a BATCH_SIZE of new
    random points,
        observing their values, and updating the current best candidate to
        the running list.
    """
    rand_x = torch.rand(BATCH_SIZE, 6)
    next_random_best = neg_hartmann6(rand_x).max().item()
    best_random.append(max(best_random[-1], next_random_best))
    return best_random
```

Now let us plot the results and analyze their performances in one place. The following code snippet generates three lines representing the average cumulative best candidates (observation) across each BO step, with the vertical bars denoting the standard deviation from the mean:

```
# plot the results
def ci(y):
    return 1.96 * y.std(axis=0) / np.sqrt(N_TRIALS)

GLOBAL_MAXIMUM = neg_hartmann6.optimal_value

iters = np.arange(N_BATCH + 1) * BATCH_SIZE
y_ei = np.asarray(best_observed_all_qei)
y_kg = np.asarray(best_observed_all_qkg)
y_rnd = np.asarray(best_random_all)
```

```
fig, ax = plt.subplots(1, 1, figsize=(8, 6))
ax.errorbar(iters, y_rnd.mean(axis=0), yerr=ci(y_rnd), label="random",
linewidth=1.5)
ax.errorbar(iters, y_ei.mean(axis=0), yerr=ci(y_ei), label="qEI",
linewidth=1.5)
ax.errorbar(iters, y_kg.mean(axis=0), yerr=ci(y_kg), label="qKG",
linewidth=1.5)
plt.plot([0, N_BATCH * BATCH_SIZE], [GLOBAL_MAXIMUM] * 2, 'k', label="true
best objective", linewidth=2)
ax.set_ylim(bottom=0.5)
ax.set(xlabel='number of observations (beyond initial points)',
ylabel='best objective value')
ax.legend(loc="lower right")
```

Running this set of code generates Figure 7-1. The result shows that the qKG acquisition function has the best performance, which is likely due to its less myopic nature that assesses the global domain instead of only focusing on the observed ones. On the other hand, qEI only dominates in the first step and starts to perform inferior to qKG due to its more myopic and restrictive nature: the best candidate must come from the observed dataset. Both qEI and qKG perform better than the random policy, showing the benefits of employing this side computation in the global optimization exercise.

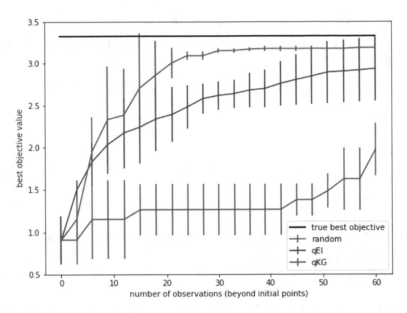

Figure 7-1. *Comparing the performance of three search policies: qEI, qKG, and random search. The qKG acquisition function has the best performance due to its less myopic nature, while qEI only dominates in the first step and starts to perform inferior to qKG due to its myopic nature. Both qEI and qKG perform better than the random policy, showing the benefits of employing this side computation in the global optimization exercise*

In the next section, we will switch gears and look at training and optimizing a convolutional neural network.

Hyperparameter Optimization for Convolutional Neural Network

A convolutional neural network (CNN) is a particular type of neural network architecture that has been shown to be an extremely powerful learner in computer vision tasks involving image data, such as object recognition. It is a sequence of layers stacked together, where each layer consists of nodes of different sizes and connects to other layers via one of the many operations: direct linear connection or nonlinear transformations such as ReLU. CNN uses the convolutional layer to learn the graphical structures of the image data (e.g., the nose is below the eyes in recognizing human faces), delivering a higher predictive accuracy than the vanilla fully connected network (FCN) yet using only a small subset of weights/parameters.

Note that these parameters are part of the CNN model and are the subject of optimization when training the CNN. In particular, the current parameters would be used to generate a set of predictions for the current batch of data and obtain a corresponding loss, a measure of the quality of fit. This completes the forward pass. In the backward pass, the loss will flow back to each previous layer by taking the partial derivative with respect to each parameter. These partial derivatives will be used to update the parameters using the well-known stochastic gradient descent (SGD) algorithm. This completes one iteration.

Understanding the inner works of CNN is not the main focus of the book. Rather, we will use CNN as an example and demonstrate how to use BO to make a good choice of the many hyperparameters. Hyperparameters are different from parameters in that they must be chosen and fixed before each round of network training starts. Examples of hyperparameters include the learning rate used in SGD updates, the number of nodes in a layer, and the number of layers in a neural network. Since the learning rate is typically the first and foremost hyperparameter to be tuned when training a modern neural network, we will focus on optimizing this hyperparameter in the following section. That is, we would like to find the best learning rate for the current network architecture that gives the highest predictive accuracy for the test set.

Let us first get familiar with the image data we will work with.

Using MNIST

The MNIST dataset (Modified National Institute of Standards and Technology database) is a database of handwritten digits widely used for experimenting, prototyping, and benchmarking machine learning and deep learning models. It consists of 60,000 training images and 10,000 testing images which are normalized and center cropped. They can be accessed via the `torch.datasets` class without separate downloading.

We will cover a step-by-step implementation of training a simple CNN model using MNIST. We start with data loading and preprocessing using the `datasets` and `dataloader` subclasses in PyTorch `torchvision` package. Using these utility functions could significantly accelerate the model development process since it offers consistent handling and preprocessing of the input data. Then, we will define the model architecture, the cost, and the optimization function. Finally, we will train the model using the training set and test its predictive performance using the test set.

Let us download the MNIST dataset using the `datasets` subpackage from `torchvision`, a powerful and convenient out-of-the-box utility for accessing popular image datasets. It stores the image samples and their corresponding labels. In the following code snippet, we first import the `datasets` class and instantiate it to download the training and test data into the Data folder. In addition, the downloaded image is transformed into a tensor object, which is a multidimensional NumPy array in PyTorch. The transformation could also include scaling the image pixels from the original range [0, 255] to a standardized scale [0.0, 1.0].

Listing 7-6. Downloading the MNIST dataset

```
# Download MNIST dataset
from torchvision import datasets
from torchvision.transforms import ToTensor
train_data = datasets.MNIST(
    root = 'data',
    train = True,
    transform = ToTensor(),
    download = True,
)
test_data = datasets.MNIST(
    root = 'data',
    train = False,
    transform = ToTensor()
)
```

We can print the objects to access their profile information, including the number of observations:

```
>>> print(train_data)
Dataset MNIST
    Number of datapoints: 60000
    Root location: data
    Split: Train
    StandardTransform
Transform: ToTensor()
>>> print(test_data)
Dataset MNIST
    Number of datapoints: 10000
```

```
      Root location: data
      Split: Test
      StandardTransform
Transform: ToTensor()
```

After downloading the data in the train_data variable, the input and output are stored in the data and targets attributes, respectively. We can examine their shape as follows:

```
>>> print(train_data.data.size())
>>> print(train_data.targets.size())
torch.Size([60000, 28, 28])
torch.Size([60000])
```

The output shows that we have 60,000 images, each having a shape of 28 by 28, corresponding to the width and height of an image. Note that a color image would have a third dimension called depth or channel, which usually assumes a value of 3. Since we are dealing with grayscale images, the extra dimension of 1 is omitted.

We can visualize a few images in the training set. The following code randomly selects 25 images and plots them on a five-by-five grid canvas:

```
# Plot multiple images
figure = plt.figure(figsize=(10, 8))
cols, rows = 5, 5
# Loop over 25 places to plot the images
for i in range(1, cols * rows + 1):
    # Generate a random index to select an image
    # The item function converts the Tensor object into a scalar value
    sample_idx = torch.randint(len(train_data), size=(1,)).item()
    # Extract the image data and target label
    img, label = train_data[sample_idx]
    figure.add_subplot(rows, cols, i)
    plt.title(label)
    plt.axis("off")
    # Squeeze the image to convert the image shape from [1,28,28]
    to [28,28]
    plt.imshow(img.squeeze(), cmap="gray")
plt.show()
```

Figure 7-2 shows the output of the 25 digits. Note that each image has a shape of 28 by 28, thus having a total of 784 pixels, that is, features per image.

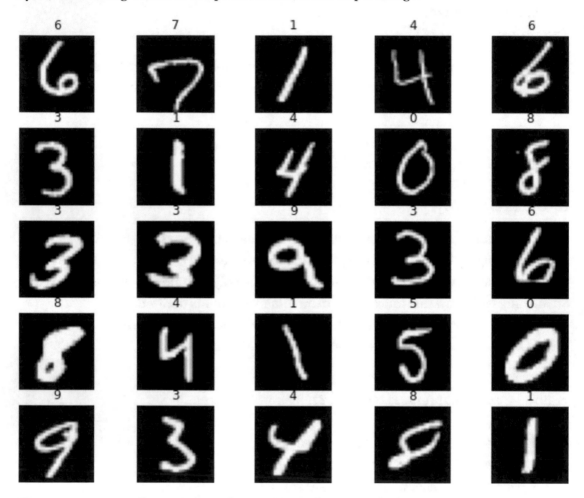

Figure 7-2. *Visualizing 25 random MNIST digits. Each digit is a grayscale image that consists of 28*28=784 pixels*

With the dataset downloaded, the next is to load them for model training purposes. The dataloader class from the torch.utils.data module provides a convenient way to load and iterate through the data in batches and perform specific transformations if needed. Passing the dataset object as an argument to dataloader enables automatic batching, sampling, shuffling, and multiprocess loading of the dataset. The following code iteratively loads the training and test set with a batch size of 100 after shuffling all the images and stores them in the loaders object:

```
# Preparing data for training with DataLoaders
from torch.utils.data import DataLoader
loaders = {
    'train' : torch.utils.data.DataLoader(train_data, # data source to be
    loaded batch_size=100, #  the number of training samples used in one
    iteration shuffle=True), # samples are shuffled and loaded in batches
    'test'  : torch.utils.data.DataLoader(test_data,
                                        batch_size=100,
                                        shuffle=True)}
```

We can observe the shape of the data loaded after iterating through loaders for the first time, as shown in the following code:

```
>>> for X, y in loaders['train']:
>>>     print("Shape of X [batch_size, channel, height, width]: ", X.shape)
>>>     print("Shape of y: ", y.shape)
>>>     break
Shape of X [batch_size, channel, height, width]:  torch.Size([100, 1, 28, 28])
Shape of y:  torch.Size([100])
```

Now let us define the CNN architecture.

Defining CNN Architecture

Our prediction model is a function that takes input data, interacts with the model parameters via a series of matrix-vector multiplication and nonlinear transformation, and outputs model predictions. In one shot, the prediction function achieves two goals: defining the model parameters and specifying the model architecture (i.e., interaction mechanism between weights and input). Achieving both objectives in one function works okay if the problem scale is small and the model is relatively simple. However, as the complexity increases in the data, the model, the cost, or the optimization procedure of the whole model training process, it is advisable to segregate the model parameters from the model architecture and modularize the respective code according to the specific purpose. In other words, the model parameters and the model architecture that specifies the interaction between the data and the parameters would have their dedicated functions.

The common practice is to use a class to wrap all these different utility functions under one roof. Specifically, we can arrange the model parameters in one function, which often manifests as the class's default __init__ function. The __init__ function contains initialized attributes when the class is first instantiated, meaning the abstract class is converted to a physical and tangible object. These attributes include essential building blocks of the class, that is, layers for the neural network and the resulting parameters implied from the layers. Instantiation is similar to buying a lego toy set with all the parts provided based on its drawing specification. These parts can also be of different sizes and structures; a big component can consist of several small ones.

In addition, the function that defines the model architecture and regulates the flow of data goes into the forward function, which serves as the instruction sheet for assembling the parts. This special function is automatically executed under the hood when the class's object is called. Thus, it acts as the default information road map that chains together the various components defined by the __init__ function, marries with the input data, specifies how they pass through the network, and returns the model prediction.

Let us look at a specific example by creating a model class called CNN, suggesting that this is a convolutional neural network. Convolution is a particular type of neural network layer that effectively processes image data. It takes the form of a small square filter (also called *kernel*) and operates on local regions of the input data by repeatedly scanning through the input surface. In the case of grayscale image data in MNIST, each image entry will manifest as a two-dimensional matrix where each cell stores a pixel value between 0 and 255.

A common practice is to follow a convolution layer with a ReLU layer (to keep positive elements only) and a pooling layer (to reduce the parameters). A pooling operation is similar to convolution in that it involves a kernel interacting with different patches of the input. However, there are no weights in the kernel. The pooling operation uses the kernel to zoom in on a specific input patch, choose a representative value such as the mean (average pooling) or the maximum (max pooling), and store it in the output feature map. Such an operation reduces the input data dimension and keeps only the (hopefully) meaningful features.

The CNN class we will create now uses all three types of layers introduced earlier. As shown in the following code listing, it inherits from the predefined base class `torch.nn.Module` which manages other backend logistics on neural network modules without having to create them from scratch. The CNN class will have two essential functions: `__init__` and `forward`. When instantiating this class into an object, say `model = CNN()`, what happens is that the model object is created using the `__init__` method of CNN. When using the model for prediction for some new data, say `model(new_data)`, we pass the `new_data` input to the `forward` function under the hood. Both functions are called implicitly.

The following code snippets show the model class. At a high level, we define two convolutional blocks `conv1` and `conv2` upon initialization in the `__init__` function. Each block sequentially applies three layers: convolution by `nn.Conv2d()`, ReLU by `nn.ReLU()`, and max pooling by `nn.MaxPool2d()`. Note that in `nn.Conv2d()`, `in_channels` refers to the number of color channels in the image data. Since we are dealing with a grayscale image, it is set to 1. The `out_channels` parameter refers to the number of kernels to be created; this will determine the depth of the resulting feature map. In other words, setting it to 16 means that we will create 16 different kernels to convolve with the input data, each learning from a different perspective and jointly forming a 16-layer feature map in the output. The `kernel_size` parameter determines the size of the square kernel. A bigger kernel will promote more feature sharing but contain more weights to tune, while a smaller kernel requires more convolution operations but can better attend to local features in the input. Finally, we have the stride parameter to control the step size when moving the kernel to convolve with the next patch, and the padding parameter to adjust for the shape of the output by adding or padding zeros to the peripheral of the input.

Listing 7-7. Defining the CNN model architecture

```
# Define the Convolutional Neural Network model class
import torch.nn as nn
class CNN(nn.Module):
    # Specify the components to be created automatically upon instantiation
    def __init__(self):
        super(CNN, self).__init__()
        # The first convolutional block
        self.conv1 = nn.Sequential(
            nn.Conv2d(
                in_channels=1,
```

```
                out_channels=16,
                kernel_size=5,
                stride=1,
                padding=2,
            ),
            nn.ReLU(),
            nn.MaxPool2d(kernel_size=2),
        )
        # The second convolutional block
        self.conv2 = nn.Sequential(
            nn.Conv2d(16, 32, 5, 1, 2),
            nn.ReLU(),
            nn.MaxPool2d(2),
        )
        # The final fully connected layer which outputs 10 classes
        self.out = nn.Linear(32 * 7 * 7, 10)
    # Specify the flow of information
    def forward(self, x):
        x = self.conv1(x)
        x = self.conv2(x)
        # Flatten the output to shape (batch_size, 32 * 7 * 7)
        x = x.view(x.size(0), -1)
        output = self.out(x)
        return output
```

Lumping together multiple simple layers in one block and repeating such blocks multiple times with different configurations is a common way of defining relatively complex neural networks. Such structure in convolutional neural networks also helps learn the compositionality of image data. A sufficiently trained neural network could extract low-level features, such as edges in the early layers, and high-level patterns, like objects in the bottom layers.

In addition, we also define a fully connected layer as the last layer of the network in the initialization function via nn.Linear(). Note that we need to pass in the correct number of nodes for the input in the first parameter and the output in the second parameter. The input size is determined based on the network configuration in previous layers, and the output is set to ten since we have ten classes of digits to classify.

And lastly, the forward function chains together these components sequentially. Pay attention when concatenating a convolutional layer with a fully connected layer. It is necessary to flatten the multidimensional cubic feature map into a single-dimensional vector for each image entry. In this case, we use the view() function to reshape the data into a single column, as indicated by the -1 parameter, while keeping the batch size (retrieved as the first dimension using x.size(0)) intact.

Besides, the training process would significantly accelerate when using a GPU. When running the accompanying notebook in Colab, a hassle-free online Jupyter notebook platform, the GPU resource can be enabled without additional charge (up to a specific limit, of course) by simply changing the hardware accelerator option in the runtime type to the GPU.

In the following code snippet, we create the model by instantiating the class into an object via model = CNN().to(device), where we use the device variable to determine the use of a GPU if available. Printing out the model object gives the specification of different components defined by the __init__ function as follows:

```
>>> device = "cuda" if torch.cuda.is_available() else "cpu"
>>> print(f"Using {device} device")
>>> model = CNN().to(device)
>>> print(model)
Using cuda device
CNN(
  (conv1): Sequential(
    (0): Conv2d(1, 16, kernel_size=(5, 5), stride=(1, 1), padding=(2, 2))
    (1): ReLU()
    (2): MaxPool2d(kernel_size=2, stride=2, padding=0, dilation=1,
    ceil_mode=False)
  )
  (conv2): Sequential(
    (0): Conv2d(16, 32, kernel_size=(5, 5), stride=(1, 1), padding=(2, 2))
    (1): ReLU()
    (2): MaxPool2d(kernel_size=2, stride=2, padding=0, dilation=1, ceil_
    mode=False)
  )
  (out): Linear(in_features=1568, out_features=10, bias=True)
)
```

When the neural network architecture starts to scale up and become complex, it is often helpful to print out the architecture for better clarification of its composition. In the following code snippet, we resort to the torchsummary package to ease the visualization task by passing in the size of an input entry. The output shows the model architecture from top to bottom, with each layer sequentially suffixed by an integer. The output shape and number of parameters in each layer are also provided. A total of (trainable) 28,938 parameters are used in the model. Note that we do not have any nontrainable parameters; this relates to the level of model fine-tuning in transfer learning.

```
from torchsummary import summary
>>> summary(model_cnn, input_size=(1, 28, 28))
----------------------------------------------------------------
        Layer (type)            Output Shape         Param #
================================================================
          Conv2d-1          [-1, 16, 28, 28]             416
            ReLU-2          [-1, 16, 28, 28]               0
       MaxPool2d-3          [-1, 16, 14, 14]               0
          Conv2d-4          [-1, 32, 14, 14]          12,832
            ReLU-5          [-1, 32, 14, 14]               0
       MaxPool2d-6            [-1, 32, 7, 7]               0
         Linear-7                  [-1, 10]          15,690
================================================================
Total params: 28,938
Trainable params: 28,938
Non-trainable params: 0
----------------------------------------------------------------
Input size (MB): 0.00
Forward/backward pass size (MB): 0.32
Params size (MB): 0.11
Estimated Total Size (MB): 0.44
----------------------------------------------------------------
```

Training CNN

The CNN model contains the preset architecture and initial weights, which will get optimized as optimization proceeds. Specifically, this optimization is guided by a direction that minimizes the loss of the training set. In the case of classifying MNIST digits, the loss function takes the form of a cross-entropy loss:

```
# Define the cost function
loss = nn.CrossEntropyLoss().cuda()
```

Recall that the optimization procedure is an iterative process. The prediction function generates an output based on the current weights in each iteration. The cost function then takes the prediction and the target label to calculate a cost. The optimization step derives the gradient of the cost for each weight and updates all the weights using the gradient descent type of algorithm. This process repeats until convergence, that is, the weights or cost remains largely the same and does not change much as the training iteration proceeds. For the actual optimization to kick in, we define the following SGD optimizer:

```
# Define the optimizer
from torch import optim
optimizer = optim.SGD(model_cnn.parameters(), lr = 0.01)
>>> optimizer
SGD (
Parameter Group 0
    dampening: 0
    differentiable: False
    foreach: None
    lr: 0.01
    maximize: False
    momentum: 0
    nesterov: False
    weight_decay: 0
)
```

We can now put together the preceding components into one entire training loop. Based on the predefined model and data loader, the following code listing defines the training function that performs SGD optimization via one complete pass through the training dataset. We also track the cost evolution as the weights get updated across different training batches, and this process is then repeated based on the specified number of epochs.

Listing 7-8. Defining the model training procedure

```
# Define the training function
def train(model, loaders, verbose=True):
    # Control the behavior of certain layers by specifying the
    training mode
    model.train()
    # Extract the total number of images to track training progress
    total_img = len(loaders['train'].dataset)
    # Extract and iterate through each batch of training data
    for batch, (X, y) in enumerate(loaders['train']):
        # Pass to GPU for faster processing
        X, y = X.to(device), y.to(device)
        # Call the forward method (under the hood) to produce prediction
        pred = model(X)
        # Calculate the current loss
        loss_fn = nn.CrossEntropyLoss().cuda()
        loss = loss_fn(pred, y)
        # Clear existing gradients
        optimizer.zero_grad()
        # Perform backpropagation and compute gradients
        loss.backward()
        # Update weights using SGD
        optimizer.step()
        # Print loss at every 100th batch; each batch has 100 image-
        label pairs
        if(verbose):
            if batch % 100 == 0:
                loss, current_img_idx = loss.item(), batch * len(X)
                print(f"loss: {loss:>7f}  [{current_img_idx:>5d}/
                {total_img:>5d}]")
```

We will also define a function to check the model's performance on the test set. The following code listing generates model predictions and calculates the test accuracy by comparing them against the target labels. Since no learning is needed, the model is set to the evaluation mode and put under the torch.no_grad() context when generating the predictions.

Listing 7-9. Defining the model testing procedure

```
# Define the test function
def test(model, loaders, verbose=True):
    # Control the behavior of certain layers by specifying the
    evaluation mode
    model.eval()
    # Extract the total number of images to in the test set
    total_img = len(loaders['test'].dataset)
    correct = 0
    # Disable gradient calculation
    with torch.no_grad():
        for X, y in loaders['test']:
            X, y = X.to(device), y.to(device)
            pred = model(X)
            # Add the correct prediction for each batch
            correct += (pred.argmax(1) == y).type(torch.float).sum().item()
            correct /= total_img

    if verbose:
        print(f"Test accuracy: {correct:>0.3f}")

    return correct
```

Now let us test both functions over three epochs. Note that each epoch means a full pass of the whole dataset, including both training and test sets:

```
num_epochs = 3
for t in range(num_epochs):
    print(f"Epoch {t+1}\n-------------------------------")
    train(model_cnn, loaders, verbose=False)
    test_accuracy = test(model_cnn, loaders, verbose=True)
>>> print("Done!")
```

```
Epoch 1
-------------------------------
Test accuracy: 0.914
Epoch 2
-------------------------------
Test accuracy: 0.937
Epoch 3
-------------------------------
Test accuracy: 0.957
Done!
```

The result shows over 95% test set accuracy in just three epochs, a decent feat given the current model architecture and learning rate. We would like to answer the following question in the following section: Which learning rate could give us the optimal test set accuracy given the current training budget?

Optimizing the Learning Rate

As the utility functions get big in both size and number, it is a good habit to contain relevant ones in a single class, making it easier for code management and abstraction. To this end, we define a class called CNN_CLASSIFIER to manage all CNN-related training and performance check. In particular, we wish to treat it as a black-box function that accepts a specific learning rate and returns the corresponding test set accuracy. This class is defined as follows:

Listing 7-10. Defining a customized module to manage model training and testing

```
from torch.autograd import Variable

class CNN_CLASSIFIER(nn.Module):
    # Specify the components to be created automatically upon instantiation
    def __init__(self, loaders, num_epochs=10, verbose_train=False,
    verbose_test=False):
        super(CNN_CLASSIFIER, self).__init__()
        self.loaders = loaders
        self.num_epochs = num_epochs
        self.model = CNN().to(device)
```

```python
        self.criterion = nn.CrossEntropyLoss().cuda()
        self.verbose_train = verbose_train
        self.verbose_test = verbose_test

    def forward(self, learning_rate):
        self.optimizer = optim.SGD(self.model.parameters(), lr =
        learning_rate)

        for t in range(self.num_epochs):
            print(f"Epoch {t+1}\n-------------------------------")
            self.train(verbose=self.verbose_train)
            test_accuracy = self.test(verbose=self.verbose_test)

        return test_accuracy

# Define the training function
def train(self, verbose=False):
    # Control the behavior of certain layers by specifying the
    training mode
    self.model.train()
    # Extract the total number of images to track training progress
    total_img = len(self.loaders['train'].dataset)
    # Extract and iterate through each batch of training data
    for batch, (X, y) in enumerate(self.loaders['train']):
        # send to GPU if available
        X, y = Variable(X).cuda(), Variable(y).cuda()
        # Call the forward method (under the hood) to produce prediction
        pred = self.model(X)
        # Calculate the current loss
        loss = self.criterion(pred, y)
        # Clear existing gradients
        self.optimizer.zero_grad()
        # Perform backpropagation and compute gradients
        loss.backward()
        # Update weights using SGD
        self.optimizer.step()
        # Print loss at every 100th batch; each batch has 100 image-
        label pairs
```

```
            if(verbose):
                if batch % 100 == 0:
                    loss, current_img_idx = loss.item(), batch * len(X)
                    print(f"loss: {loss:>7f}  [{current_img_idx:>5d}/
                    {total_img:>5d}]")
        return self

    # Define the test function
    def test(self, verbose=False):
        # Control the behavior of certain layers by specifying the
        evaluation mode
        self.model.eval()
        # Extract the total number of images to in the test set
        total_img = len(self.loaders['test'].dataset)
        correct = 0
        # Disable gradient calculation
        with torch.no_grad():
            for X, y in self.loaders['test']:
                X, y = X.to(device), y.to(device)
                pred = self.model(X)
                # Add the correct prediction for each batch
                correct += (pred.argmax(1) == y).type(torch.float).
                sum().item()
        correct /= total_img

        if verbose:
            print(f"Test accuracy: {correct:>0.3f}")

        return correct
```

One particular point to pay attention to is the way we initialize the SGD optimizer. In the forward() function, we pass self.model.parameters() to optim.SGD() to show that we are optimizing the parameters of the model within this class, given the learning rate input. The model will not get updated if such awareness is removed, that is, not accessing the model parameters as part of the self attributes.

We can now perform the same trial as before, in a much more convenient manner:

```
nn_classifier = CNN_CLASSIFIER(loaders=loaders, num_epochs=3, verbose_
test=True)
>>> nn_classifier(learning_rate=0.01)
Epoch 1
--------------------------------
Test accuracy: 0.902
Epoch 2
--------------------------------
Test accuracy: 0.938
Epoch 3
--------------------------------
Test accuracy: 0.953
0.9531
```

Entering the Full BO Loop

As before, we will first generate a set of initial conditions. Since each trial requires training a full CNN model using the training set and testing on the test set, we need to be mindful of the potential runtime. Running more epochs will definitely give us a more accurate estimate of the (real) observation, that is, the test set accuracy for the specific learning rate, although at the cost of more computing time. We limit the training budget to three epochs for all runs in the following exercise.

The following function generate_initial_data_cnn() gives us a similar function to generate the initial conditions as before, with two notable differences. The first difference is that our random samples are now uniformly sampled from a lower bound 0.0001 and an upper bound 10. The second difference is that all relevant objects now live in the GPU and typed double, achieved by .type(torch.DoubleTensor).cuda(). In addition, note that we set the noise level to zero as there is no randomness in the observations.

```
# generate initial dataset
NOISE_SE = 0

def generate_initial_data_cnn(n=5):
    # generate training data
    # train_x = torch.rand(n, 1, device=device, dtype=dtype)
```

```
train_x = torch.distributions.uniform.Uniform(0.0001,10).sample([n,1]).
type(torch.DoubleTensor).cuda()
train_y = []
for tmp_x in train_x:
    print(f"\nCurrent learning rate: {tmp_x.item()}")
    nn_classifier = CNN_CLASSIFIER(loaders=loaders, num_epochs=3,
    verbose_test=True)
    tmp_y = nn_classifier(learning_rate=tmp_x.item())
    train_y.append(tmp_y)

train_y = torch.tensor([[tmp] for tmp in train_y]).type(torch.
DoubleTensor).cuda()
best_observed_value = train_y.max().item()
return train_x, train_y, best_observed_value
```

Now let us generate three initial observations. The result from running the following code shows a highest test set accuracy of 10%, not a good start. We will see how to obtain more meaningful improvements later.

```
train_x, train_y, best_observed_value = generate_initial_data_cnn(n=3)
>>> print(train_x)
>>> print(train_y)
>>> print(best_observed_value)
tensor([[6.0103],
        [2.8878],
        [5.0346]], device='cuda:0', dtype=torch.float64)
        tensor([[0.1032],
        [0.1010],
        [0.0980]], device='cuda:0', dtype=torch.float64)
        0.10320000350475311
```

After initializing and updating the GP model, let us take qEI as the example acquisition function to obtain the next learning rate. The following function `optimize_acqf_and_get_observation_cnn()` helps us achieve this:

Listing 7-11. Obtaining the next learning rate by optimizing the acquisition function

```
from botorch.optim import optimize_acqf
bounds = torch.tensor([[0.0001], [10.0]], device=device, dtype=dtype)
BATCH_SIZE = 3
NUM_RESTARTS = 10
RAW_SAMPLES = 512

def optimize_acqf_and_get_observation_cnn(acq_func):
    """Optimizes the acquisition function, and returns a new candidate and
    a noisy observation."""
    # optimize
    candidates, _ = optimize_acqf(
        acq_function=acq_func,
        bounds=bounds,
        q=BATCH_SIZE,
        num_restarts=NUM_RESTARTS,
        raw_samples=RAW_SAMPLES,  # used for initialization heuristic
        options={"batch_limit": 5, "maxiter": 200},
    )

    # observe new values
    new_x = candidates.detach()
    # exact_obj = neg_hartmann6(new_x).unsqueeze(-1)  # add output dimension
    # new_y = exact_obj + NOISE_SE * torch.randn_like(exact_obj)

    new_y = []
    for tmp_x in new_x:
        print(f"\nCurrent learning rate: {tmp_x.item()}")
        nn_classifier = CNN_CLASSIFIER(loaders=loaders, num_epochs=3,
        verbose_test=True)
        tmp_y = nn_classifier(learning_rate=tmp_x.item())
        new_y.append(tmp_y)

    new_y = torch.tensor([[tmp] for tmp in new_y]).type(torch.
    DoubleTensor).cuda()

    return new_x, new_y
```

217

Note that we use a for loop to obtain each of the three learning rates in the next stage. This could be further parallelized if we can access multiple GPUs and design the corresponding search strategy.

Let us test this function to obtain the next three learning rates as well:

```
new_x_qei, new_y_qei = optimize_acqf_and_get_observation_cnn(qEI)
>>> print(new_x_qei)
>>> print(new_y_qei)
tensor([[ 0.1603],
        [10.0000],
        [ 7.9186]], device='cuda:0', dtype=torch.float64)
tensor([[0.9875],
        [0.0958],
        [0.1010]], device='cuda:0', dtype=torch.float64)
```

Similarly, we would edit the function to generate the learning rates using the random search strategy. This function, update_random_observations_cnn(), is defined as follows:

Listing 7-12. Randomly choosing learning rates as benchmark for comparison

```
def update_random_observations_cnn(best_random):
    """Simulates a random policy by drawing a BATCH_SIZE of new
    random points,
        observing their values, and updating the current best candidate to
        the running list.
    """

    rand_x = torch.distributions.uniform.Uniform(0.0001,10).sample
    ([BATCH_SIZE,1]).type(torch.DoubleTensor).cuda()
    rand_y = []
    for tmp_x in rand_x:
        print(f"\nCurrent learning rate: {tmp_x.item()}")
        nn_classifier = CNN_CLASSIFIER(loaders=loaders, num_epochs=3,
        verbose_test=True)
        tmp_y = nn_classifier(learning_rate=tmp_x.item())
        rand_y.append(tmp_y)
```

```
rand_y = torch.tensor([[tmp] for tmp in rand_y]).type(torch.
DoubleTensor).cuda()
next_random_best = rand_y.max().item()
best_random.append(max(best_random[-1], next_random_best))

return best_random
```

Finally, we are ready to serve the main course. The following code snippet introduces the same three competing search strategies: qEI, qKG, and random policy, all of which proceed for a total of ten steps by setting N_BATCH=10. Note that the total running time takes around one hour in a single GPU instance, so grab a cup of coffee after you execute the accompanying notebook!

Listing 7-13. The full BO loop for three competing hyperparameter search strategies

```
# number of runs to assess std of different BO loops
N_TRIALS = 3
# indicator to print diagnostics
verbose = True
# number of steps in the outer BO loop
N_BATCH = 10
best_observed_all_qei, best_observed_all_qkg, best_random_all = [], [], []

# average over multiple trials
for trial in range(1, N_TRIALS + 1):

    print(f"\nTrial {trial:>2} of {N_TRIALS} ", end="")
    best_observed_qei, best_observed_qkg, best_random = [], [], []

    # call helper functions to generate initial training data and
    initialize model
    train_x_qei, train_y_qei, best_observed_value_qei = generate_initial_
    data_cnn(n=3)
    mll_qei, model_qei = initialize_model(train_x_qei, train_y_qei)

    train_x_qkg, train_y_qkg = train_x_qei, train_y_qei
    best_observed_value_qkg = best_observed_value_qei
    mll_qkg, model_qkg = initialize_model(train_x_qkg, train_y_qkg)
```

```python
    best_observed_qei.append(best_observed_value_qei)
    best_observed_qkg.append(best_observed_value_qkg)
    best_random.append(best_observed_value_qei)

    t0_total = time.monotonic()

    # run N_BATCH rounds of BayesOpt after the initial random batch
    for iteration in range(1, N_BATCH + 1):
        t0 = time.monotonic()

        # fit the models
        fit_gpytorch_mll(mll_qei.cpu());
        mll_qei = mll_qei.to(train_x)
        model_qei = model_qei.to(train_x)

        fit_gpytorch_mll(mll_qkg.cpu());
        mll_qkg = mll_qkg.to(train_x)
        model_qkg = model_qkg.to(train_x)

        # define the qEI acquisition function using a QMC sampler
        qmc_sampler = SobolQMCNormalSampler(sample_shape=torch.Size([MC_
        SAMPLES]))

        # for best_f, we use the best observed noisy values as an
        approximation
        qEI = qExpectedImprovement(
            model = model_qei,
            best_f = train_y.max(),
            sampler = qmc_sampler
        )

        qKG = qKnowledgeGradient(model_qkg, num_fantasies=MC_SAMPLES)

        # optimize and get new observation
        new_x_qei, new_y_qei = optimize_acqf_and_get_observation_cnn(qEI)
        new_x_qkg, new_y_qkg = optimize_acqf_and_get_observation_cnn(qKG)

        # update training points
        train_x_qei = torch.cat([train_x_qei, new_x_qei])
        train_y_qei = torch.cat([train_y_qei, new_y_qei])
```

```
    train_x_qkg = torch.cat([train_x_qkg, new_x_qkg])
    train_y_qkg = torch.cat([train_y_qkg, new_y_qkg])

    # update progress
    best_random = update_random_observations_cnn(best_random)
    best_value_qei = max(best_observed_qei[-1], new_y_qei.max().item())
    best_value_qkg = max(best_observed_qkg[-1], new_y_qkg.max().item())
    best_observed_qei.append(best_value_qei)
    best_observed_qkg.append(best_value_qkg)

    # reinitialize the models so they are ready for fitting on next
    iteration
    mll_qei, model_qei = initialize_model(
        train_x_qei,
        train_y_qei
    )
    mll_qkg, model_qkg = initialize_model(
        train_x_qkg,
        train_y_qkg
    )

    t1 = time.monotonic()

    if verbose:
        print(
            f"\nBatch {iteration:>2}: best_value (random, qEI, qKG) = "
            f"({max(best_random):>4.2f}, {best_value_qei:>4.2f}, {best_
            value_qkg:>4.2f}),"
            f"time = {t1-t0:>4.2f}.", end=""
        )
    else:
        print(".", end="")

best_observed_all_qei.append(best_observed_qei)
best_observed_all_qkg.append(best_observed_qkg)
best_random_all.append(best_random)

t1_total = time.monotonic()
print(f"total time = {t1_total-t0_total:>4.2f}.")
```

We plot the cumulative best-observed test set accuracies of the three search strategies as shown in Figure 7-3. Note that the standard deviation is very large due to insufficient trials in each experiment, thus exceeding the maximum value (i.e., 100%) and giving only an indicative result.

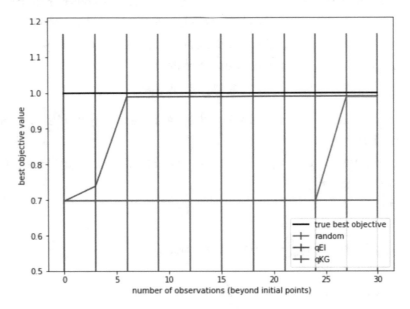

Figure 7-3. *Comparing the performance of three search policies in identifying the optimal learning rate. The qEI acquisition function located the highest-performing learning rate in just two steps, while qKG exhibited more variation across all steps. The random strategy only caught up in the last few steps*

For the full scale of implementation covered, please visit the accompanying notebook at `https://github.com/Apress/Bayesian-optimization/blob/main/Chapter_7.ipynb`.

Summary

In this chapter, we introduced the process of optimizing in a full BO loop, covering a synthetic case using Hartmann and a practical hyperparameter tuning case using CNN. Specifically, we covered the following:

- The full BO loop consists of generating a set of initial conditions, updating the GP posterior, initializing the acquisition function, and proposing the next sampling location to complete one step, repeated for multiple steps.

- qKG is a better-performing strategy than qEI when seeking the global optimum of the Hartmann function, while qEI performs the best when optimizing the learning rate of CNN. Both perform better than the random search strategy.

- Parameters in a neural network are optimized using stochastic gradient descent, while its hyperparameters are optimized using BO.

Index

A

Acquisition functions, 29–30, 70, 72–74, 77–86, 98, 131, 140, 145, 148–149, 151–159, 185, 190–192

Adam optimizer, 116

AdditiveKernel() function, 127

Airline count dataset, 124–125

Airline passenger count, 124–129

Analytic EI in BoTorch
  best_value, 135
  ExpectedImprovement class, 135, 136
  GP surrogate with optimized hyperparameters, 134, 135
  Hartmann function, 132, 133
  inner optimization routine, 140–148
  optimization, 137–140
  to() method, 137

Approximate dynamic programming (ADP) methods, 82

Automatic differentiation, 102, 146

Ax platform, 185

B

backward() function, 104

batch_initial_conditions variable, 141, 145

Bayesian decision theory
  Bellman's principle of optimality, 79–82
  multi-step lookahead policy, 76–79
  optimal policy, 72–74

utility-driven optimization, 74, 76

Bayesian formula, 11, 16

Bayesian inference
  actual observation, 24
  aims, 11
  Bayesian formula, 11
  Bayes' rule, 23, 24
  Bayes' theorem, 23
  conjugate, 13
  denominator, 13
  normal distribution, 13, 25
  observation model, 13
  posterior distribution, 12, 24, 25
  prior distribution, 12
  prior uniform distribution, 13
  probability density function, 23
  subjective expectation, 24

Bayesian optimization
  aims, 1
  BoTorch, 185
  characteristics, 31
  definition, 1
  global optimization, 1, 31
  interactions, 1
  libraries, 96–98
  loop, 30–31, 73, 74, 92, 96, 98, 104, 135, 187, 192–198, 215–222
  objective function, 1
  overall process, 2
  statistics (*see* Bayesian statistics)
  surrogate function, 185

N

O

Printed in the United States
by Baker & Taylor Publisher Services